Penguin Nature Guides

Fungi of
Northern Europe · 1

Larger Fungi (excluding Gill-Fungi)

Sven Nilsson and Olle Persson
Illustrated by Bo Mossberg

Translated from the Swedish by David Rush
Edited and adapted by Dr David Pegler and
Brian Spooner

Penguin Books

Penguin Books Ltd, Harmondsworth,
Middlesex, England
Penguin Books, 625 Madison Avenue,
New York, New York 10022, U.S.A.
Penguin Books Australia Ltd, Ringwood,
Victoria, Australia
Penguin Books Canada Ltd, 2801 John Street,
Markham, Ontario, Canada L3R 1B4
Penguin Books (N.Z.) Ltd, 182–190 Wairau Road,
Auckland 10, New Zealand

Svampar i naturen 1 first published by
Wahlström & Widstrand 1977
This translation published 1978

Printed in Portugal by Gris Impressores
Filmset in Monophoto Times by
Northumberland Press Ltd, Gateshead, Tyne and Wear

Contents

Symbols

☉   edible.

(☉)   edible after parboiling, soaking, drying etc.

⊗   inedible, suspect or inferior.

✠   poisonous.

✠✠   very poisonous.

✠(☉)   poisonous but edible after parboiling, soaking, drying etc.

      The symbol following the heading for a group covers all species mentioned under that heading, unless otherwise shown.

Foreword

Under the old firs the fir-cones lie in their hundreds, their scales wide open. Only fir-cones? No, some large fine morels are growing there too. In the aspen grove nearby wood anemones are blooming.

The sun dapples the ground through the green of the hazel trees. Ivy creeps along the ground, and climbs with thick, snake-like branches around the tree-trunks. Amidst the ivy, sorrel and mosses, Golden Russulas are shining gold and red.

It is late autumn – grey damp weather. The lime trees in the avenue are nearly bare of leaves. The yellowing husks of fruit still hang there. The lichen-covered trunks are dark with moisture. On the bark grow tiny delicate Mycenas.

These are glimpses of the scenes that have inspired me in my work, and that I have tried to pass on in my pictures of fungi and their surroundings.

Bo Mossberg

The big, fleshy, edible mushrooms that we find in autumn are only part of an enormous family of fungi which embraces a host of very different species, many of them insignificant or invisible to the naked eye. Some are illustrated and described in the introductory chapter, but this book concentrates on the large fleshy fungi. In the British Isles there are at least 10,000 different types of fungus. Of these about two thousand are conspicuous, and can be described as the larger fungi. Many of these are illustrated here. They are grouped according to our current understanding of their relationship. We have also tried to give a thorough description, both in words and pictures, of how and where the various species grow – a valuable guide in searching for and identifying fungi.

This volume deals with the groups of larger fungi which have no gills, also the Roll-rims (*Paxillus*) and the Spike-caps (*Gomphidius*) which we think of as being closely related to the boleti. Volume 2 deals with the agarics. We have mainly followed the system used in Morten Lange's *Systematisk Botanik, Svampe* (Copenhagen, 1955).

We have tried to use acceptable common names. The name of the author is also given with the Latin names. We must become familiar with the Latin names if we wish to extend our knowledge by reading foreign works about fungi. The purpose of the book, in fact, is not only to provide an introduction to the world of the fungi, but also to whet the reader's appetite for a wider knowledge of the subject.

Great emphasis is placed on the illustrations, which in many cases are reproduced life-size. Some fungi are shown at several stages of development, in different forms, and with variations of colour. If a species is included in the book, it should be possible to identify it from the pictures, but the text goes into considerable detail and important ecological facts are also noted. Background details in the pictures are very important and in certain cases we have illustrated the whole botanical environment. We have not felt able, however, to provide a 'key' to the various species until the book has been tested in practice.

Although the book results from a close collaboration between the three of us, we are individually responsible for various sections. The layout and illustrations are the work of Bo Mossberg, and are based mainly on sketches of fresh fungi. The general introductions in both volumes were written by Sven Nilsson. Olle Persson is responsible for the texts on individual species (except for the polypores) in Volume 1, and in Volume 2 for the texts on the species of *Hygrophorus, Lepiota, Agaricus, Cortinarius* and *Lactarius*. The other texts in Volume 2, and the text on the polypores in Volume 1 are, with a few exceptions, by Sven Nilsson.

We are most grateful to many specialists on the various groups of fungi for their opinions and advice, and for the identification of sketches. In particular we wish to mention Aino Käärik (Stockholm), John Axel Nannfeldt (Uppsala), Ronald H. Petersen (Knoxville, Tennessee, U.S.A.) and Tauno Ulvinen (Uleåborg, Finland). Rolf Lidberg (Sundsvall) has generously put at our disposal his knowledge of the fungi of northern Sweden, and has also contributed material for sketches.

We want especially to thank Nils Suber, the father-figure of Swedish experts on our subject. For many years he has unstintingly shared his knowledge. Through him – and also through the late Seth Lundell of Uppsala – we have been able to share in a Swedish tradition of research that goes back to Elias Fries.

Bo Mossberg   *Sven Nilsson*   *Olle Persson*

What is a fungus?

Nature is dominated by green plants. A green world is a condition of our existence. The green substance chlorophyll is decisively important in the process of photosynthesis. This means that green plants create, from water and carbon dioxide and with the help of light, their own nourishment: organic compounds rich in energy. In the green world there are also many equally important organisms which have no chlorophyll. Among these are the fungi, mainly seen for a month or two in the autumn.

Fungi cannot create their own nourishment. Like animals they must live as saprophytes on dead organic material, or as parasites on other living organisms. Tradition divides the organic world into the animal and the plant kingdoms, with fungi included in the latter. But fungi are so unlike other organisms, mainly in their cell formation and in the way they absorb nourishment, that they should really be placed in their own group or kingdom.

Between 50,000 and 100,000 species are known, and probably almost as many are yet to be discovered. They, and the bodies through which they spread and multiply – their spores – are found throughout every type of environment on earth.

The Greeks called fungi *mykes*, a word also used for objects shaped like a mushroom. The word survives in the name of the ancient Greek city Mycenae. The city was founded by Perseus on the spot, according to some legends, where he happened to lose the cap of his scabbard and, according to others, where he found a mushroom that quenched his thirst. The word mycology, the study of fungi, also comes from the Greek.

When we talk about mushrooms or toadstools, most people think of the boletes, russulas and chanterelles found in autumn. These, however, are merely the fruit-bodies of a fairly small group of fungi. Most fungi are microscopic. The fungus itself consists of finely ramifying threads, called hyphae, which form a network, the mycelium. In certain lower fungi the threads are simply forked outgrowths of plasma. In others they are divided into cells by cross-walls (septa) of a more or less complex structure.

Even in the pre-Christian era chronicles and legends told of fungi and their effects, how they sprang up and then vanished, and their curious manner of growth. The deadly effect of certain poisonous fungi caused them to be feared and respected. In later times, even scientists did not understand their classification. Linnaeus placed them in the genus *Chaos* under *Vermes* (worms). It was the Swede, Elias Fries (1794–1878), the father of mycology, who laid the foundations for detailed research into fungi, and for the general and scientific interest in them.

Man also learned early to value fungi as a food delicacy. Even if we have now realized their significance and have put them to industrial and medical use, they are still surrounded by a certain mystique. Most people still know little about them and how they grow.

The structure and development of fungi

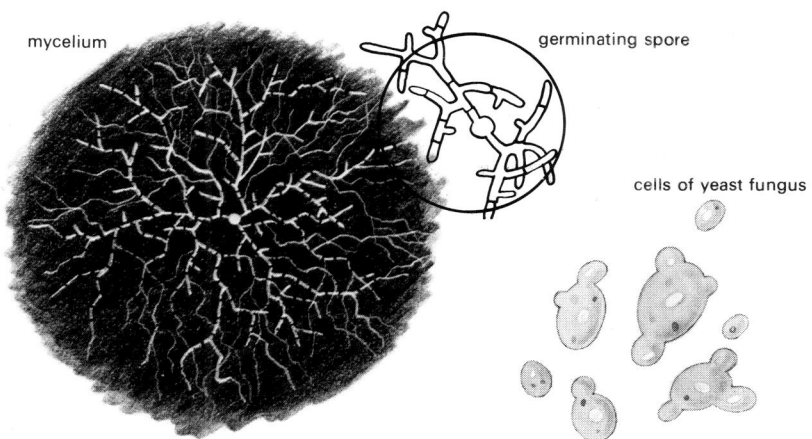

mycelium

germinating spore

cells of yeast fungus

The threads, or hyphae, grow out radially from the spore and form a constantly branching network. Most fungi are made up of hyphae which combine to form more or less compact bodies of tissue. A number of fungi are single-celled, such as the yeast fungi. New cells develop by division (see illustration)

What we usually notice are the fruitbodies of fungi, like this Penny Bun Bolete. The fruitbodies vary greatly in type and size. Although fungi contain no chlorophyll, and some are colourless, many have beautifully coloured fruitbodies

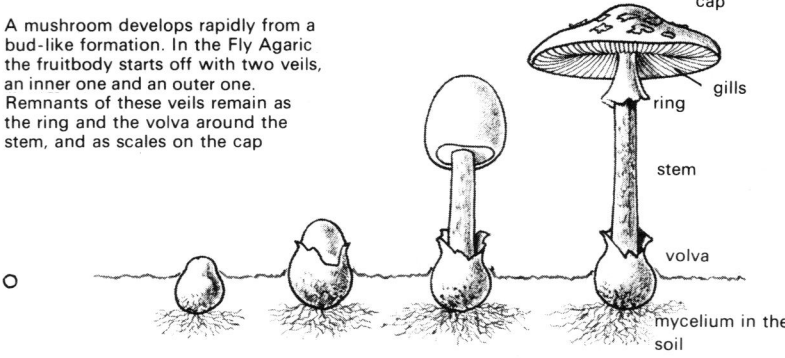

A mushroom develops rapidly from a bud-like formation. In the Fly Agaric the fruitbody starts off with two veils, an inner one and an outer one. Remnants of these veils remain as the ring and the volva around the stem, and as scales on the cap

cap

gills

ring

stem

volva

mycelium in the soil

The hyphae from two spores of different sexes come together and form a mycelium whose cells have two nuclei. This binucleate mycelium then produces microscopic club-shaped formations, the basidia, on which the spores are formed. In the case of an agaric the basidia can be seen on the surface of the gills. The spores give the gills their special colouring

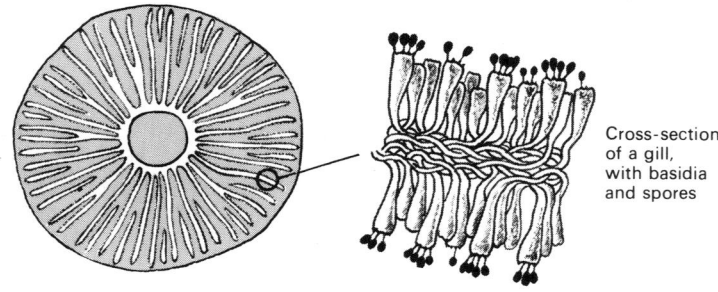

Cross-section of a gill, with basidia and spores

Spores on the surface of the gill

Release of spores

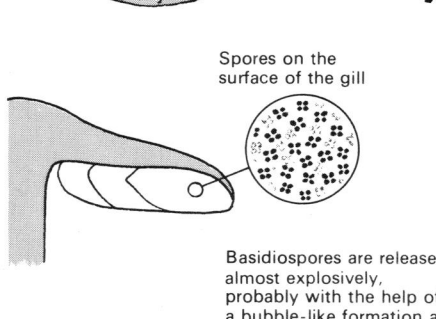

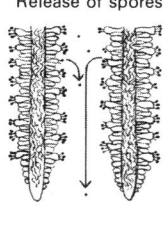

Basidiospores are released almost explosively, probably with the help of a bubble-like formation at the base of the spore

basidiospore

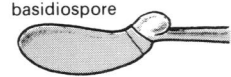

The basidiomycetes

The greater part of a fungus consists of mycelium, which occurs either in the ground or in organic matter, both dead and living. This mycelium, whose cell-walls unlike those of green plants consist of chitin instead of cellulose, has grown from germinating spores. The mycelium may be short-lived and reduced or it may have a life of several years and be widely distributed over and within the substratum. It is from the mycelium that the spore-producing part of the fungus is formed – the fruitbody. The appearance of this fruitbody, and the way it forms its spores, is our basis for the systematic division of fungi into various groups.

Let us look first at an agaric's structure and the way it forms its fruitbody. A fungus has two or more different types of mycelium – two sexes if you like – although we cannot in fact see this. These two sexes are often assigned the symbols + and − When two compatible types of mycelium come together, a binucleate mycelium is formed, in which each cell has nuclei from both types. This mycelium forms the spore-producing parts of the fungus. The fruitbody first appears as a small bud on the mycelium, and then gradually assumes its characteristic shape.

The young fruitbodies are often surrounded at first by one or more 'veils'. In the Fly Agaric there is an outer (or universal) veil, which bursts when the stem lengthens. Fragments of the veil remain in the form of scales on the cap of the fungus and a volva or 'stocking' around the stem. An inner veil covers the young gills. This comes away, and remains in the mature fungus only as a ring on the stem. You can see a typical loose ring of this kind on Parasol mushrooms. In species of *Cortinarius* the remains of the veil form a net like a spider's web on the cap, the gills and the stem, whereas in *Gomphidius* species they form a slimy covering.

On the underside of the cap of the full-grown fungus there are gills with a fruiting layer (hymenium) of spore-producing organs or basidia. These are club-like swellings at the ends of the hyphae. Each basidium normally forms four spores, set on horn-like protuberances or sterigmata. After the two different types of nuclei have fused, four nuclei are gradually formed by division, each ending up in its own spore.

The full-grown spores, which are often distinctively coloured (see first section of vol. 2), are shot out with a small explosion and then fall vertically between the gills. The gills of an agaric are always vertically arranged, so that the spores can be effectively released.

The fruitbody of an agaric can form very quickly. We say that things 'spring up like mushrooms' when they arrive quickly, unexpectedly and in large numbers. A fruitbody, as it grows, can exert enormous strength. The Ink Cap, for example, can force its way up through the asphalt surfaces of roads and the cement floors of buildings.

From a classification point of view the agarics belong to the family of basidial fungi or basidiomycetes. A typical and mature agaric consists of a cap, with gills on the underside, and a stem. Both the cap and the stem are formed from a compact mass of hyphae. At the base of the stem you can observe some loose threads of mycelium, parts of the mycelium which grows underground in the soil. There are many other kinds of basidial fungi, in which the basidia are not formed on gills, but in tubes, on spines and on ridges. Sometimes these fungi cannot be divided into cap and stem (see p. 56). In the simplest kinds the basidia grow scattered on the mycelium, but more often they form some kind of cohesive layer, a fruit layer or hymenium, either over or within a fruitbody.

One of the largest fruitbodies known for fungi with caps is the Australian *Boletus portentosus* whose cap is nearly half a metre wide and which weighs several kilograms. Only the Giant Puffball (see p. 52) and certain polypores have larger fruitbodies.

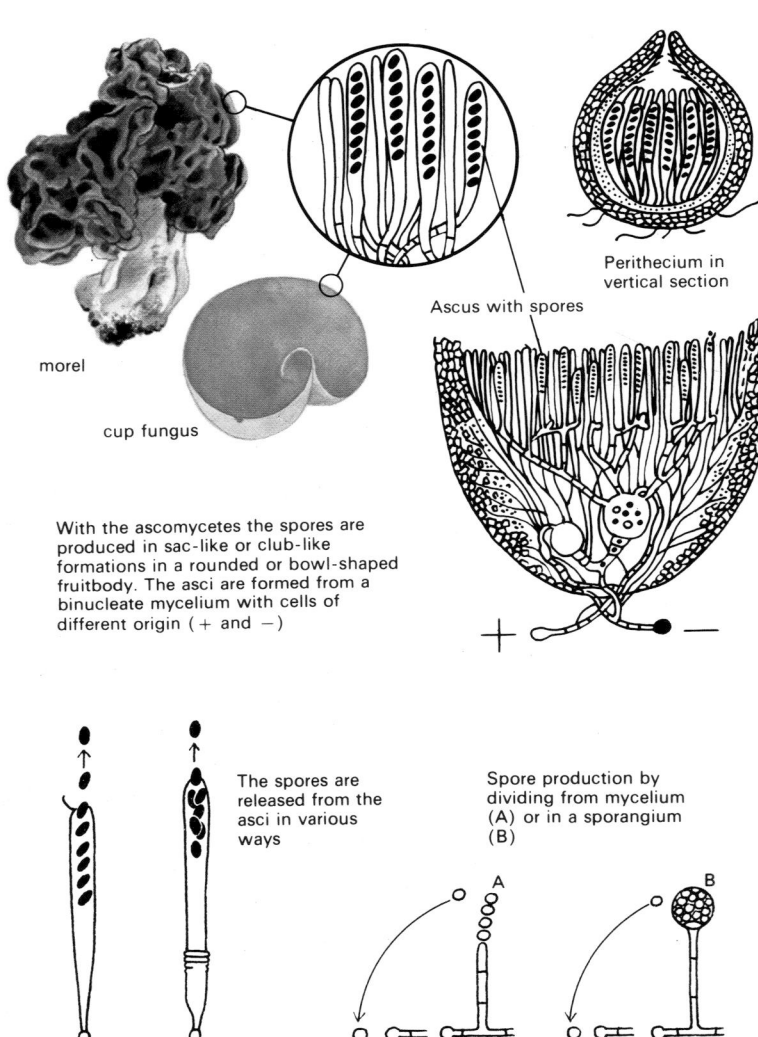

morel

cup fungus

Perithecium in vertical section

Ascus with spores

With the ascomycetes the spores are produced in sac-like or club-like formations in a rounded or bowl-shaped fruitbody. The asci are formed from a binucleate mycelium with cells of different origin (+ and −)

The spores are released from the asci in various ways

Spore production by dividing from mycelium (A) or in a sporangium (B)

Ascomycetes and lower fungi

The cup fungi or ascomycetes have different types of fruitbodies from the agarics. These consist of cup-shaped or more or less closed forms to which the hymenium is attached, either inside or out. With these fungi, too, there are genetically different types of spores and mycelium. With many of them, under suitable conditions, two compatible kinds of mycelium form club-like 'sexual organs', which then unite.

10

Nuclei from the male organ travel across to the female one. From the latter the hyphae then grow, whose cells are double-sexed like those of the basidial fungi. Clubs or sacs, the asci (singular, ascus), grow from the ends of the hyphae. In each of these the two nuclei mingle and become one. This nucleus, which therefore has two sets of chromosomes, is then divided and reduced to four nuclei. After a further splitting the ascus finally contains eight nuclei. Gradually eight spores are formed, each with a nucleus. These spores, ascospores, which may have one cell or several, are released in various ways from the asci. Electron microscope investigations have shown that the upper part of the ascus has a complex structure. In certain species the ascus has a kind of lid, in others an expanding pore. Hyphae that do not form asci form the fruitbody. This may be cup-shaped and known as an apothecium as, for example, in the case of the cup fungi. The morels have a fruitbody of this kind in an altered form (see p. 34). A large group of ascomycetes have rounded fruitbodies, often dark coloured and more or less closed. Called perithecia, these are normally microscopic. Sometimes the fruitbodies are sunk in a tightly-knit tissue, or stroma, as with the 'cushion fungi' (see p. 44).

The simplest ascomycetes do not form a fruitbody. With the yeast fungi the asci consist of free cells. With other groups the asci are scattered throughout the mycelium. The types that form fruitbodies have their asci in a fruit-layer or hymenium, either on the fruitbody or within it.

With the less advanced fungi, the thread fungi or phycomycetes, the spores are normally formed in simple spore-producing organs or sporangia. This happens asexually, without any previous union between mycelia or nuclei that are genetically different. Many aquatic fungi of the thread type have mobile spores, zoospores, equipped with 'whiplashes' or flagella. There is also a sexual phase involving sexual organs, in which male cells or nuclei are carried over to female cells or nuclei.

In most types of fungi, spores can also be formed asexually and more or less directly from the mycelium. These spores, normally called conidia, split off directly from the mycelium or from special spore-bearing branches of the mycelium. They may have one cell or several. Sometimes the spore-bearing hyphae close together to form a simple type of fruitbody. It has been possible to connect many of these conidial forms with definite species of ascal or basidial fungi. In other cases only the conidial stage is known. Brush-moulds, for example, are the conidial stages of certain ascomycetes.

From the process by which the spores are formed it is clear that ascomycetes, like the basidiomycetes, have two stages or generations: a stage where there is a single set of chromosomes in the cell-nuclei, and a stage where there is a double one. However, this alternation of generations is quite unlike the one that occurs in mosses and ferns.

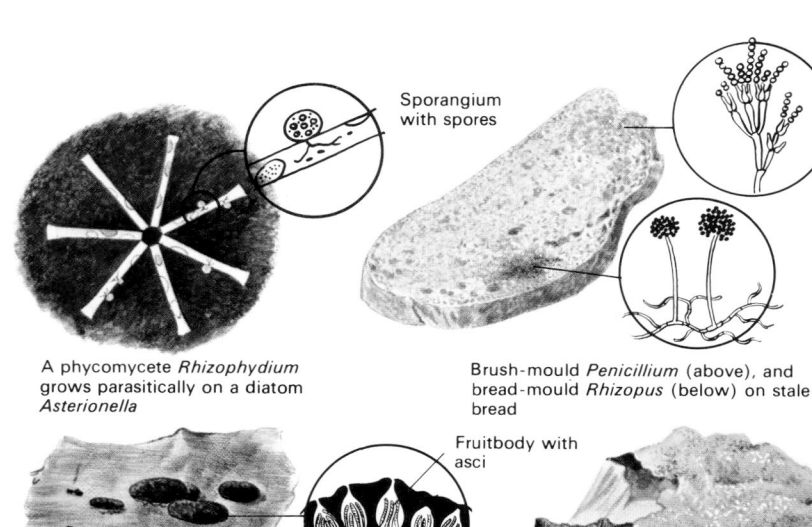

A phycomycete *Rhizophydium* grows parasitically on a diatom *Asterionella*

Sporangium with spores

Brush-mould *Penicillium* (above), and bread-mould *Rhizopus* (below) on stale bread

Fruitbody with asci

stroma

With the 'cushion fungus' *Hypoxylon* the fruitbodies are sunk in a dense, partly dead tissue, the stroma.
'Flowers of Tan', *Fuligo* (yellow) and 'Stump Lycogala' *Lycogala* (pale red) are two very common slime fungi

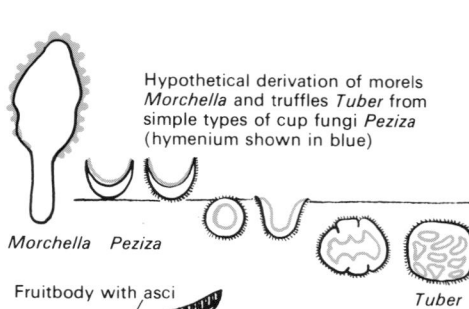

Hypothetical derivation of morels *Morchella* and truffles *Tuber* from simple types of cup fungi *Peziza* (hymenium shown in blue)

Peziza   *Helvella*   *Morchella*   *Peziza*

Tuber

Fruitbody with asci

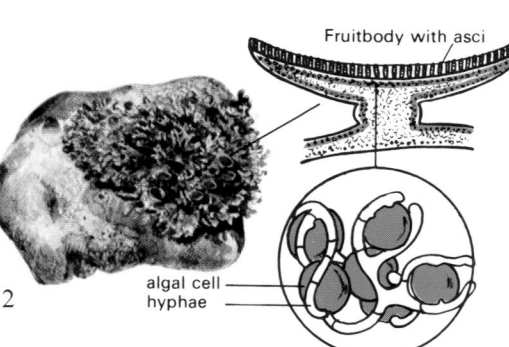

The body of the lichens consists of fungal hyphae and a layer of algal cells (shown in green). The fruitbody is normally cup-shaped, as with the lichen shown here, *Parmelia saxatilis*

algal cell
hyphae

Various types of fungi

The simplest fungi are single-celled microscopic organisms. Others are very complex in formation and striking in appearance. Fungi can be divided into three main groups. The lowest varieties form a heterogeneous group generally known as the thread fungi (phycomycetes). Their hyphae have no dividing walls and they never form complex fruitbodies. Many of them are aquatic, living on or in dead organic matter in the water. An example of a thread fungus that lives on land is the so-called 'false mildew'. It causes disease in many useful plants. The saprophytic moulds are common and familiar. Some of them, such as bread-mould, are often seen on stale food.

The other two main groups are the ascomycetes and basidiomycetes. The most important difference between them, as we have already seen, is the way the spores are produced.

The majority of ascomycetes are small, and can only be studied under a magnifying glass or microscope. But some are large and attractively coloured. The simplest are the yeast fungi and their relatives, which have no fruitbodies. The remainder can be split into three groups. The first group have fruitbodies which consist only of fairly irregular balls of hyphae. To this group belong the fruiting stages of brush-moulds. With the second group the asci are formed in a cohesive stroma (see p. 11). But the great majority of ascomycetes belong to the third group in which asci form a cohesive hymenium in or on a well-developed fruitbody. We can distinguish two main types from the appearance of the fruitbody. The cup fungi or Discomycetes have apothecia – fruitbodies shaped more or less like plates or cups. The spores can be shot out freely and spread by the wind. Morels and truffles are variant forms of cup fungi. One can think of a morel as a large cup or plate which has been twisted in and out. The hymenium thus covers the whole of the fungus's wrinkled surface. Several types of morel are more or less cup-shaped, or begin their development as cup-shaped structures. The picture on the opposite page shows how the morel species are believed to have developed. Underground truffles can be seen to derive from cup-shaped fungi in the same way, and look like apothecia which have been crumpled together. In this case, however, the spores cannot be spread freely. Instead, the whole fruitbody is eaten by an animal and the spores pass through the animal's gut. This process is in fact necessary to enable the spores to germinate.

The flask fungi or pyrenomycetes have a rounded fruitbody or perithecium. It is often microscopic. Sometimes, however, large numbers of fruitbodies are formed together in or on a stroma. These stromata are clearly visible as brownish warts or cushion-like formations.

Most lichen fungi also belong to the cup fungi. Lichens are double organisms, consisting of algae and fungi. It is a very stable form of co-existence or symbiosis. A few lichen fungi belong to the basidiomycetes.

The organisms known as slime fungi (myxomycetes) are normally included as members of the fungal kingdom or as being closely related to it. They are regarded as very primitive, with some animal as well as fungal features. They have an animal-like stage when they are mobile lumps of protoplasm (plasmodia): under certain conditions they develop a fungus-like spore-producing stage, a sporangium. Two common types are illustrated.

The third principal group of fungi are the basidiomycetes. The only important difference between these and the ascomycetes is, as we have seen, the way the spores

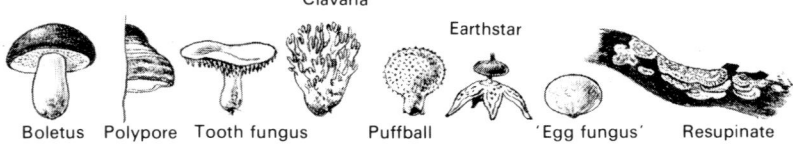

Clavaria

Boletus Polypore Tooth fungus Puffball Earthstar 'Egg fungus' Resupinate

The basidiomycetes have many types of fruitbody, adapted to different kinds of existence. We see here the common and typical 'mushroom-shaped' fungi with caps; the 'stomach fungi', often strangely shaped, and resupinate fungi whose fruitbodies consist only of a thin coating, or cushion-like jellyish formations, on branches and trunks. The rusts and smuts are other greatly modified basidiomycetes

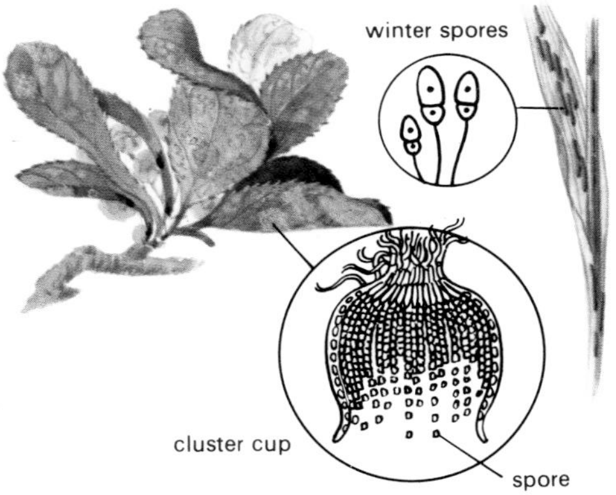

winter spores

cluster cup

spore

The fruitbodies of Black Rust appear on the barberry as orange patches, either round or running together. On wheat the patches are orange to very dark brown. They represent different stages of the same fungus

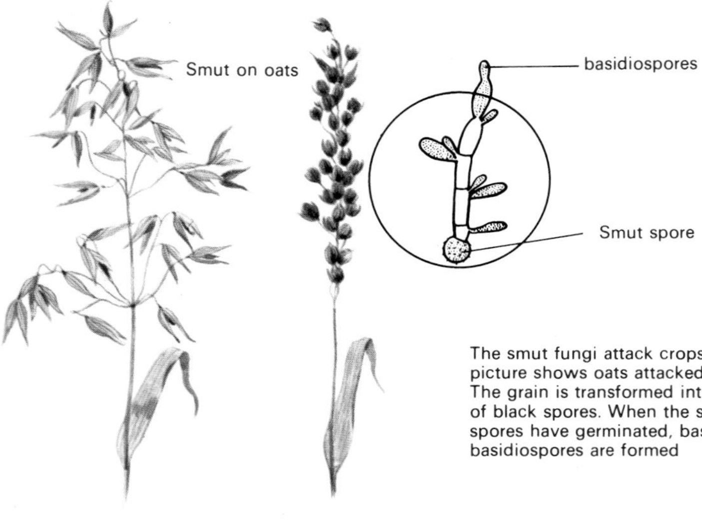

Smut on oats

basidiospores

Smut spore

The smut fungi attack crops. The picture shows oats attacked by smut. The grain is transformed into a mass of black spores. When the smut spores have germinated, basidia and basidiospores are formed

14

are formed. With certain basidiomcytes 'clamps' occur on the binucleate mycelium. These are curious bow-shaped connections over the cross-walls between the cells. If there are no such 'clamps', it is hard to tell whether individual mycelium or hyphae come from a basidiomycete or from an ascomycete.

Basidiomycetes are a very heterogeneous group, and researchers are still divided about the relationship and classification of the various types. They can be broadly divided into two groups. To one group belong the agarics, polypores, clavarias, puff-balls and various others. All of these, except the puffballs and their relatives have basidia on an external hymenium. Thus the spores can be spread directly. These fungi are therefore normally called hymenomycetes. In the case of the puffballs and related species, the basidia and spores are inside the cavities of the fruitbody, and the spores are set free only when the fruitbody is opened in some way. In this group, called stomach fungi or gasteromycetes, there are many strangely-shaped species, such as earthstars and Bird's Nests. Many tropical species have very beautiful shapes and colours.

The other large group of basidiomycetes consists of the jelly fungi and the rusts and smuts. With this group the formation of the basidia differs from that described for the agarics (p. 9). The basidia look different and the basidiospores often do not germinate directly, but form buds of spores which then germinate. Jelly fungi spread out their more or less slimy fruitbodies on dead branches, trunks and stumps. They are wart-like or irregularly wrinkled with the consistency of jelly or cartilage.

The rusts and blights are much reduced parasitic fungi. The rusts are firmly linked to certain plant species and cannot be cultivated on an artificial medium. Some have different host plants at different stages of their development. One example is Black Rust (*Puccinia graminis*) which switches between barberry and certain grasses, including wheat. If we followed the life-cycle of Black Rust on barberry and wheat we would find that the basidiospores germinate on a barberry leaf. Gradually, ball-shaped microscopic formations appear on the upper side of the leaf, and spores from genetically different formations mingle and produce a binucleate mycelium. This in turn gives rise to bowl-shaped spore-producing organs or 'cluster cups' – orange-red specks on the underside of the leaf. Spores from these are spread to wheat, where a new type of organ in the fungus creates 'summer spores', which can spread the rust to new wheat plants. In late summer the fungus on the wheat forms two-celled 'winter spores'. These survive the winter, and grow into simple basidia, from which new spores then split off.

In our climate Black Rust flourishes by switching hosts between barberry and grasses, and therefore the barberry is controlled by law. The ravages of rust fungi can be easily seen on many plants, both wild and cultivated. The cluster cup stage is clearly visible in the form of reddish specks and streaks on leaves and stalks.

The smuts also switch hosts between certain plants, but can also live partly sapro-phytically. A smut spore corresponds to the winter spore of Black Rust. It germinates into a basidium, producing buds of basidiospores, which then multiply by fresh budding. The union of genetically different spores forms a binucleate mycelium, which infects the host plant in the flower or bud. Spore colonies are gradually built up, usually in the flowering parts of the plant. When corn is attacked by smut, the grains are transformed into a mass of smut spores.

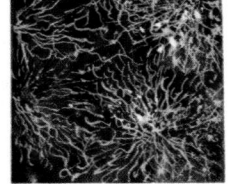

mycelium
in the soil

aquatic fungi

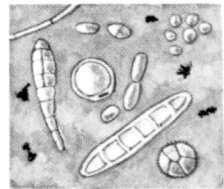

spores in the air

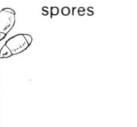

spores

threadworm

mycelium with traps

All organic material can be attacked by fungi. On dead bits of wood the grey or dark brown patches are caused by one of the world's commonest fungi *Cladosporium* (above). Fungi attack foodstuffs, vegetables and fruit. They also live as parasites on other organisms, including other fungi. An unusual type lives in soil and captures threadworms. They encircle their victims and suck nourishment from them

Fungi in the wild

Fungi are present everywhere. Their spores have even been found at a height of several thousand metres, caught in special traps on aeroplanes. Others float in the sea and other waters, and grow on nearly everything that exists in water. Fungi form a large part of the world of organisms, although we cannot see them with the naked eye. Their mycelium weaves itself into the soil, and through dead plants and animals; it breaks down dead matter, it lives as a parasite on or within other organisms, or lives in partnership with them. Fungi can attack food, clothes, timber and buildings – virtually anything. When man creates new environments, or substrata, fungi spring up, apparently adapted to just that substratum. Even the structure and mechanism of jet planes are not immune. The habitats of many fungi are highly specialized. One species can exist only on the stamens of a particular flower; another only on a particular limb of a particular species of insect.

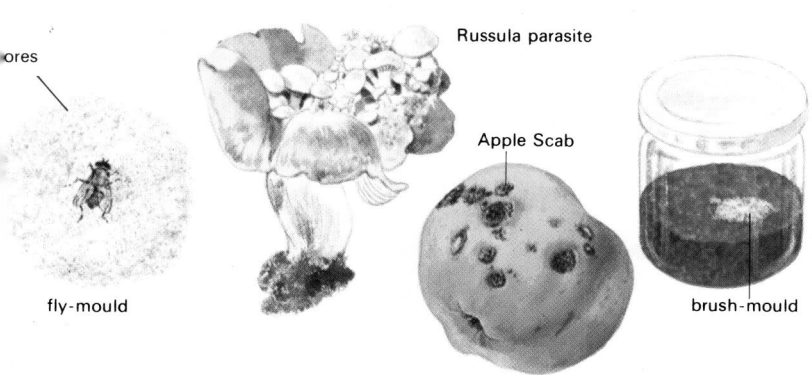

ores

fly-mould

Russula parasite

Apple Scab

brush-mould

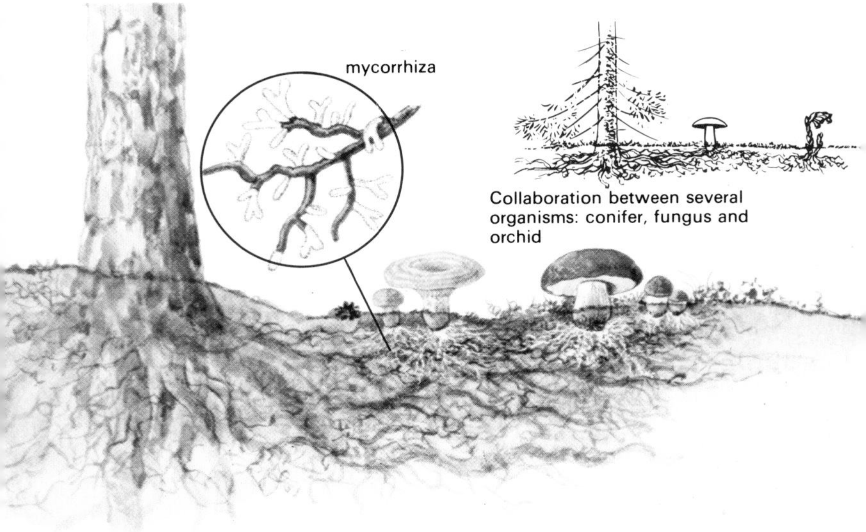

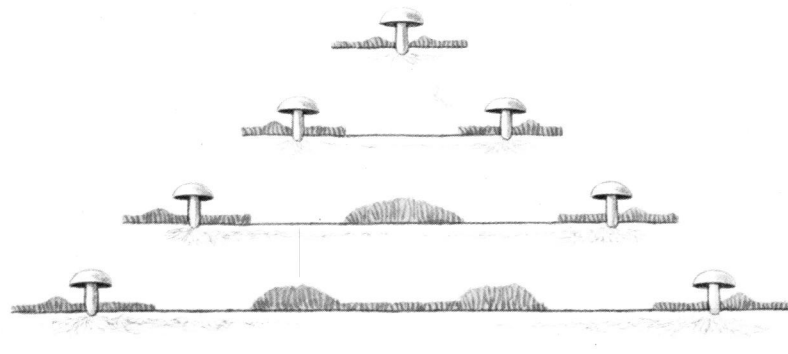

mycorrhiza

Collaboration between several organisms: conifer, fungus and orchid

Milk caps, boleti and russulas are found under many trees, forming mycorrhizal associations with them. Lift up the green covering of moss and you will see the tree roots and the fungal mycelium woven together. Note that the tiny tips of the roots are swathed in hyphae. Sometimes there are forms of co-existence involving several different organisms. Certain parasitic plants without chlorophyll are connected to trees through the mycelium of fungi, and can thus draw nourishment from the tree via the fungus

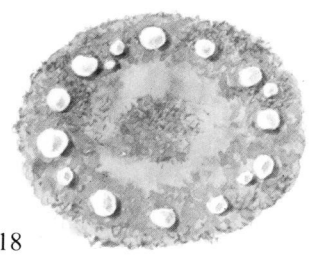

The development of a fairy ring. The mycelium grows radially, and the fruitbodies are formed on the outer edge of the circle. The vegetation becomes richer because of the nitrogen released when fungi break down dead matter

18

How fungi live

Fungi, together with bacteria, are essential to life: they break down organic matter, prevent dead matter from accumulating and bring back nourishment to the soil. The organisms that break down the remains of dead animals and plants to their component elements are called saprophytes. Many of the soil-growing fungi are saprophytic, forming what are called fairy rings – rings of fruitbodies. At one time these rings were objects of awe and superstition. People believed that the fungi sprang up where fairies danced, lightning struck or animals grazed in a ring and manured the earth.

The mycelium grows radially, along the outer edge of the ring where the fruitbodies are formed. The activity of the mycelium makes the soil rich in nitrogen and the vegetation flourishes, taking on a deeper green than its surroundings. Even when the fruitbodies have vanished dark rings show where the fungi have been. The growths of some very large fairy rings can be traced back hundreds of years.

Another group of fungi are called mycorrhizal fungi (from the Greek *mykes*, fungus, and *rhiza*, root). Many plants, mostly woodland trees, live in partnership, or symbiosis, with fungi. The mycelium of the fungus surrounds the tips of their roots, and an exchange of nourishment takes place. The fungus gets carbohydrates from the tree, and the tree thrives because the mycelium makes it easier to take in food from the soil. This is why fungi are an important aspect of forestry.

Another kind of mycorrhiza occurs with herbaceous plants – orchids, for example. The hyphae of the fungi weave themselves through the delicate roots of the plants (endomycorrhiza). The Honey Fungus is one of the fungi which forms mycorrhiza with orchids. The lichens (see p. 13) represent a more stable type of partnership – a double organism consisting of fungus and alga.

A large number of fungi are parasites, and some do great harm. We deal with some of these in another chapter. There are also other organisms which live parasitically on fungi. The grubs of many insects, the fungus gnat for example, live in the tissue of boleti and agarics. Many small animals, such as snails, damage the fruitbodies of fungi.

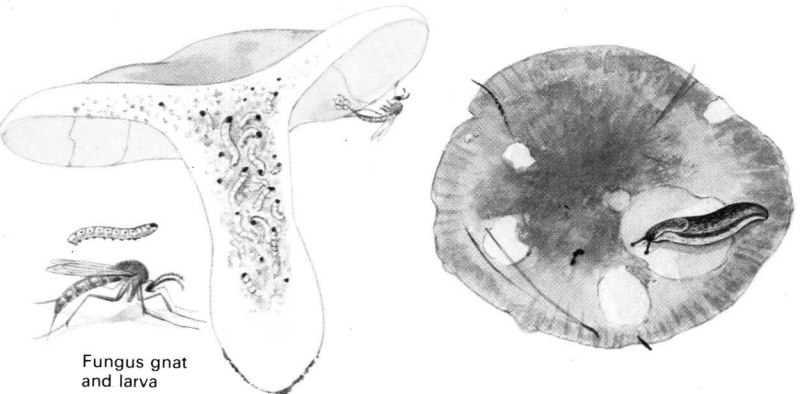

Fungus gnat
and larva

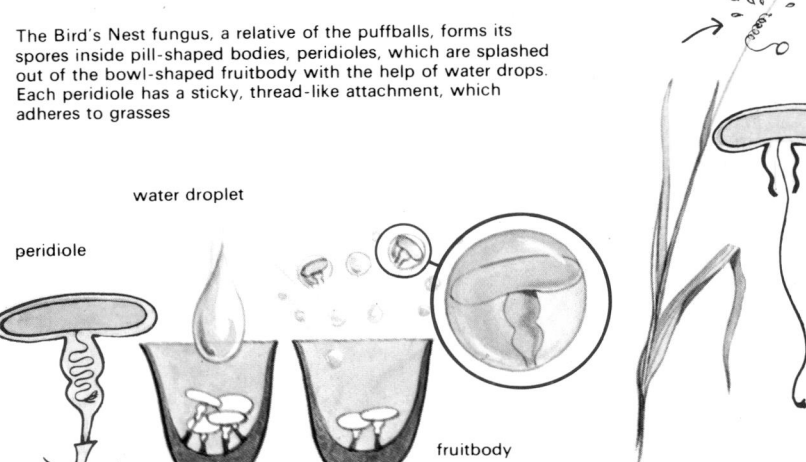

The Bird's Nest fungus, a relative of the puffballs, forms its spores inside pill-shaped bodies, peridioles, which are splashed out of the bowl-shaped fruitbody with the help of water drops. Each peridiole has a sticky, thread-like attachment, which adheres to grasses

water droplet

peridiole

fruitbody

Propagation

Fungus spores are a most effective means of propagation because they are produced in great quantity and easily spread by wind, water and other organisms. A polypore can produce millions of spores a minute, and spore formation continues for a long time. Many fungi form fruitbodies only under certain conditions or at a certain time of the year, and a large number have refined, often peculiar methods of spreading their spores. Some, like *Pilobolus*, literally explode theirs. In the common agarics and boleti there is a kind of shooting out when the spores fall straight downwards to be wafted by air currents. Aquatic fungi spores often have slimy appendages for swimming. The four-armed (tetraradiate) and S-shaped (sigmoid) spores seem to be the most common and most effective. The moulds form incredible quantities of dry light spores which are easily spread by the wind. They grow quickly on a suitable substratum such as man's food. Many ascomycetes, like those that live on manure, also spread their spores explosively. The sticky spores adhere to grasses and other plants, which are spread further when eaten by animals.

Pilobolus

How spores are spread by a cup fungus and an agaric

Tetraradiate spores of aquatic fungi

sporangium

light

spore

Growing seasons

If a fungus is to grow and form fruitbodies, it needs a suitable habitat, climate and the right degree of damp. Many fungi are found throughout the year but the agarics and boleti are autumn-growing. Alternating dry and wet weather combined with warmth are the best conditions under which fruitbodies are produced. After dry summers the fungi may not appear at all, or, as after the dry summers of 1969 and 1976, may not appear until late September. In the wet summer of 1960 the fungi came up thick and fast at the end of July and beginning of August.

Some of the earliest spring fungi are the morels. In Britain they can appear as early as April. During the summer the fairy-ring champignons, russulas, boleti and edible mushrooms start to spring up – usually in that order. Russulas are at their best at the end of August and the beginning of September.

Many species can survive a mild frost, and in mild winters you can pick mushrooms in Britain until Christmas. Velvet Shank comes up after the first frosts, and Sulphur Tuft, like a number of other woodland species, endures cold quite well. During the winter you can find polypores and jelly fungi.

Many fungi form fruitbodies only periodically. It may be several years before fruitbodies of the same fungus appear again in the same place.

Many fungi used for food have a short growing season. The Penny Bun Bolete, for instance, has a growing season of approximately two weeks in any one place, but other types may spring up and vanish within twenty-four hours. The end of the growing season in one area may be the beginning of the season in another which has a slightly different climate. Therefore it is not always possible to know when a certain species will appear. Two fungi with fairly long lives are the Shaggy Ink Cap, which can be found from midsummer to late autumn, and the Sulphur Tuft (not edible). Fungi that grow on wood can often be found in the same place over quite a long period.

Mushroom-pickers may try to draw up a schedule of where and when various fungi occur during the year but this is not always possible. Climate and moisture are often the deciding factors. In a dry year look for fungi in shady mossy places facing north. In wet summers, these areas are too moist, so hunt on heathlands, sandy soil and rocky ground.

Harmful fungi

Among organisms as widespread and which take so many forms as fungi, there are bound to be a number of species harmful to man. Many live as parasites or saprophytes on other organisms, or on matter which is either useful to man or necessary to his existence. We are therefore always waging an unequal battle with many fungal diseases.

Fungi proliferate in late summer and autumn when the green plants begin to lose their vitality and powers of resistance. Mildew, rust, smut and other forms of mould can be seen nearly everywhere. In natural woodland the ravages of mould and fungi

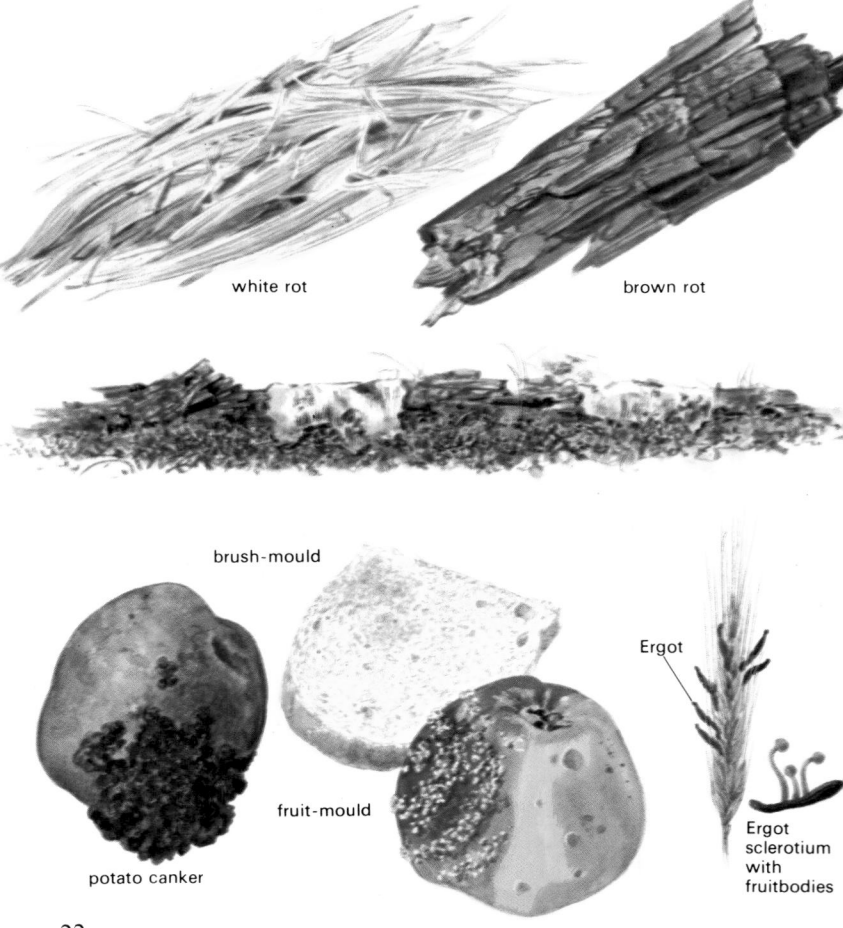

white rot

brown rot

brush-mould

Ergot

fruit-mould

Ergot sclerotium with fruitbodies

potato canker

are clearly seen. In the northern and temperate regions, where forestry is economically important, the harmful Honey Fungus and a host of similar types, especially the polypores, cause severe damage to living trees and timber. Wood used by fungi for food becomes rotten. The two kinds of rot are known by their colour – white and brown. White rot fungi are usually described as lignin-destroying, and brown rot fungi as cellulose-destroying. More accurate terms would be shrinking rots, stain rots, hole rots and white rots. Shrinking rots are brown. The wood becomes brittle and splits into cube-shaped pieces. Stain rots and hole rots cause patches of damage in the branches. The wood becomes pale, consisting only of cellulose. Finally, cavities may appear in the whole branch or it will split off. White rots occur primarily in deciduous trees, turning the wood completely white, and soft and spongy when damp. One of the white rot fungi is the Honey Fungus (see vol. 2).

Some of the more harmful fungi are the rusts and the smuts (see p. 15) which we try to control by disinfecting seeds and by spraying, but the spores can travel great distances. Special stations have been established in Europe and throughout the world to send out warnings when large quantities of spores are being carried by air currents from infected areas. One problem in combating these fungi – Black Rust, for example – is their ability constantly to form new strains that are resistant to our present methods of fighting them.

Various types of mildew, fruit-mould, leaf-mould and canker-producing fungi are harmful, and attack fruit, vegetables and root-crops on a large scale. Badly tended fruit is easily attacked by fruit-moulds such as Apple mould (*Monilia fructigena*). The mycelium and spores of the fungus form rings on the outside of the fruit. Another type of fungus found on apples is Apple Scab (*Fusicladium dendriticum*). Various kinds of fungi cause cankers on potatoes and swedes, among other crops. The Potato Canker (*Synchytrium endobioticum*), an alga-like fungus, is an example.

The mildew fungi are ascomycetes. They are highly specialized physiologically and normally bound to specific host plants. In late autumn many leaves and stalks look as if they have been dusted with a greyish powder, or covered with a thin, fluffy, fabric-like spider's web. The fluff is the mycelium of the fungus and the powder is the spores from its conidial stage. Sallow, oak and maple are particularly vulnerable to mildew. Maple Mildew (*Uncinula bicornis*) is often seen at the roadside.

On birches one sometimes sees formations rather like magpies' nests. These are caused by a relative of the yeast fungi, Witch's Broom (*Taphrina betulina*). The action of the fungus causes an abnormal development of small shoots.

One very dangerous kind of fungus is Ergot (*Claviceps purpurea*) which attacks grasses, including cereals, and especially rye. The fungus transforms the ovaries into a hard, blackish, banana-shaped formation, the sclerotium or Ergot. This passes the winter on the ground. The following year pinkish, head-like stromata develop, with perithecia sunk in the heads. The spores from these then spread and attack new rye plants. Ergot contains powerful poisons, among them ergotamin. If anyone eats bread that has been baked with grain containing Ergot, these poisons are absorbed and cause muscular contractions in the blood-vessels. In severe cases there are cramps and burning sensations in the fingers and toes, which may wither and fall off. Sometimes arms and legs are lost in this way. The connection between the fungus and the severe illness was not understood for centuries and epidemics of Ergot poisoning raged throughout Europe. The last outbreak in Britain occurred in Manchester in 1928. Far into the twentieth century France was the major area where the disease struck. Ergotamin has, however, been used in some medicines.

Several other fungi have also proved damaging, especially when an area is wholly dependent on one or two crops that are attacked by disease. The potato grower's worst enemy is potato blight (*Phytophthora infestans*), a thread fungus related to the so-called false mildew. In the middle of the nineteenth century this fungus caused

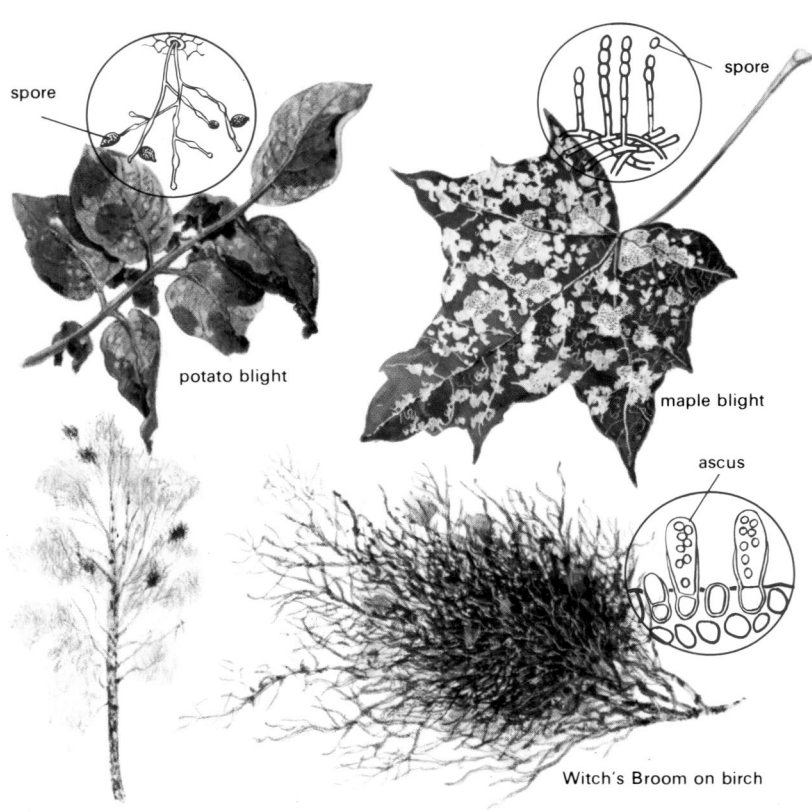

spore

spore

potato blight

maple blight

ascus

Witch's Broom on birch

great damage throughout Europe, mainly in Ireland where the population lived almost entirely on potatoes. When potato blight broke out, starvation and diseases caused by malnutrition caused a million deaths, and forced a large part of the remaining population to emigrate.

Among the familiar fungi are the moulds. Millions of pounds worth of food is discarded every day because it has been attacked by mould, and food storage is a great problem in tropical countries. Diseases caused by fungi are one of the great hindrances in combating hunger in the poorer countries.

The ubiquitous 'brush-mould' fungi (the *Penicillium* family) are found on practically everything – bread, juice, jam, fruit, vegetables and clothing. In the damp climate of the tropics these fungi are particularly damaging, and soon attack all organic matter. Probably we shall never rid ourselves of them; their spores are in the air we breathe, and on and in everything around us.

24

Another kind of fungus which is very common on dead or dying vegetation is 'grey mould' (*Botrytis cinerea*), greyish-brown spots of dusty spores on leaves and stalks, easily seen on withering plants during autumn. There are also species of the *Cladosporium* family – dark brown streaks and spots on stalks and leaves.

A number of fungi cause diseases in man. These are particularly common in the tropics, but they also exist in Europe. Mainly they cause various kinds of skin diseases. Many animals, too, are vulnerable to fungal infections, but this can be put to good use, for example, in combating harmful insects which have become more and more resistant to insecticides. These insects are a threat to useful organisms and even to man. We are constantly looking for other means of combating them, and intensive research is being done on 'biological control': causing one organism to destroy another. If a population of, say, harmful insects could be infected with a fungal disease, then it might be possible to destroy them. The fungi which attack nematode worms are another example of fungi that can be used in the fight against organisms harmful to man.

Useful fungi

Among the many thousands of species of fungi a large number are directly beneficial to man in food, drink, medicines and industry.

The yeast fungi were among the first important organisms in man's daily life. Even in prehistoric cultures it was known how to bake bread, brew beer and make wine by using the processes which take place through the 'breathing' of fungi. Single-celled fungi increase rapidly by budding when they are grown in surroundings that contain sugar and are rich in oxygen. Through the breathing process the sugar is broken down, and carbon dioxide and water are formed as energy develops. In the absence of oxygen fermentation anaerobic breathing takes place, in which the sugar is broken down to alcohol and carbon dioxide. In baking, the development of carbon dioxide makes the dough light, and the bread porous and appetizing. Compressed yeast consists of cells of the yeast fungus (*Saccharomyces cerevisiae*) which have been pressed together.

Beer is made with rye malt, mixed with warm water. The sprouting grain contains enzymes which break down the grain's starch into sugar. The yeast cannot itself break down the starch. When spices are added the sugary liquid – the wort – is boiled, and the yeast is added. The yeast transforms the wort into beer.

The wine-yeast fungus (*Saccharomyces ellipsoideus*) grows wild in the ground, easily attaches itself to grapes and thus rapidly increases in the fermenting tub.

wine-yeast fungus

Several kinds of yeast fungi
(*Saccharomyces*) are used in baking, and in
the making of wine and beer.
The wine-yeast fungus has a large number
of strains which makes the difference in taste
of wines from various districts

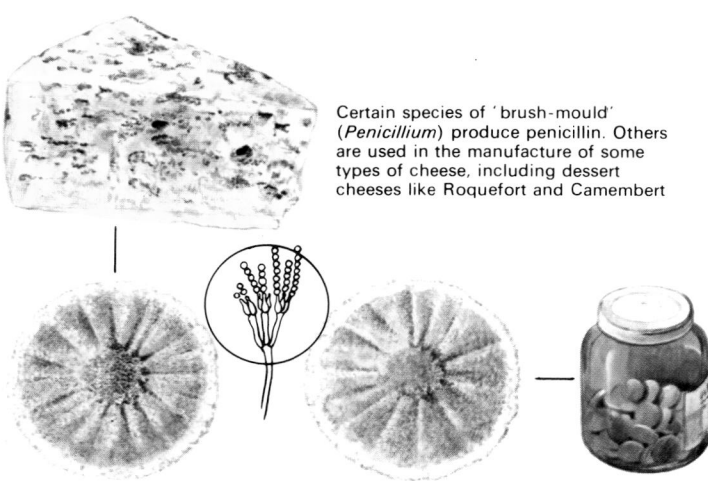

Certain species of 'brush-mould'
(*Penicillium*) produce penicillin. Others
are used in the manufacture of some
types of cheese, including dessert
cheeses like Roquefort and Camembert

Cultures of brush-mould *Penicillium*

The famous Japanese rice-wine, sake, has been made for thousands of years with the help of fungi of the genus *Aspergillus*. The fungi produce the enzyme diastase, which can break down starch into fermentable kinds of sugar. Citric acid is another product obtained by cultivating the common and well-known fungus *Aspergillus niger* in a nutrient solution.

Moulds are used in some foods regarded as delicacies. Several species of brush-mould (*Penicillium*) are used in the production of cheese. The famous French Roquefort was first manufactured by accident. People in that region stored their sheep's-milk cheese in limestone caves. A certain type of brush-mould, for which the cheese provided suitable nourishment, flourished in this moist cool environment. The mould later came to be known as *Penicillium roqueforti*. The cheese, which was rotting through the activity of the fungus, proved to be a delicacy. Various kinds of 'veined cheese' are now produced commercially, and their manufacture is a science in itself. This rotting process is called a 'maturing process'. The special taste and consistency of Camembert depends on two fungi: the ' brush-mould ' *Penicillium camemberti* and a conidial fungus *Oidium lactis* commonly found in milk.

The ' brush-moulds ', among them *Penicillium notatum*, also produce one of our most important medicines, penicillin. Its antibiotic effect was discovered by chance in 1929 by Sir Alexander Fleming, and today there are various types. In its manufacture the fungi are grown in a flowing sterilized solution. The penicillin produced by the fungus is then separated from the liquid and purified.

Fungi, or the effects of fungal activity, are used in the manufacture of many foodstuffs, medicines and other important products.

In western Europe fungi are not regarded as a real foodstuff, or even as a complement to basic foods. They are more a delicacy or a flavouring. Gathering and preparing one's own mushrooms is seen more as a hobby than as an addition to the diet. But in eastern and parts of central Europe fresh and preserved fungi are an important foodstuff, and have probably been so for hundreds of years. In ancient Greece and Rome mushrooms were a highly-prized dish. The Emperor Mushroom (*Amanita caesarea*), an orange-red fungus which the Romans called 'Boletus', was and still is one of the most highly-regarded fungi in southern Europe.

Since the seventeenth century France has been the major European country regularly to cultivate fungi, and to use them in cooking. Mushroom-eating came to Britain with other French customs but mainly as a delicacy. In the late nineteenth century and during the two world wars, attempts were made to popularize mushrooms in western Europe as a substitute for meat, but they never quite caught on. In central Europe they are very popular and in eastern Europe mushrooms are gathered in great quantities for eating or preserving. In Finland between one and two million kilos of mushrooms are harvested every year. An army of trained pickers deliver them to a chain of buyers, who sell them for tinning, pickling and drying.

The commercial cultivation of mushrooms has expanded rapidly since the eighteenth century, especially in France. Large quantities are grown in abandoned stone quarries in the districts around Paris under carefully controlled moisture and temperature conditions. The mushrooms are grown in beds consisting of baked horse manure or synthetic compost. The process is begun with mycelium, grown in a laboratory, which is mixed with the compost: this usually takes place in the dark. British mushroom-eaters usually want the white 'button' variety but in other countries the fruitbodies are allowed to develop. The flavour lies largely in the gills, and is strongest in full-grown mushrooms. Today anyone can grow various kinds of mushroom on a small scale, but not the kinds that form mycorrhiza, such as the Penny Bun Bolete, Chanterelle and Saffron Milk Cap. In several countries, especially Japan, the Padi Straw mushroom is grown on sawdust in glass containers or plastic bags. Materials for growing the common mushrooms at home are widely available from seed merchants and laboratories.

The highly-prized true truffle is grown outdoors, especially in the French province of Périgord. It is cultivated in conjunction with the oak tree, with which it forms mycorrhiza. In the French countryside it was – and still is – a picturesque sight to see truffles being gathered with the help of pigs.

The Japanese have long cultivated a fungus called shiitake (*shiia* is a kind of tree, and *take* means fungus). Holes are bored in logs of hornbeam or oak and filled with mycelium. The logs are then leant against trees in woods where they are kept constantly damp. The fungi then grow out of the logs. The black jelly mushroom, an important ingredient in Chinese cooking, is the Jew's Ear (*Hirneola auricula-judae*) or one of its relatives.

The commercial cultivation of mycelium to provide cheap protein food has been tried in many countries. In Bulgaria, for example, there are now some foods which

27

'Pigs rooting for truffles'. A drawing from a French postcard

contain a certain percentage of mycelium. These experiments are mainly being made with wood fungi, especially with the mycelium of the Scaly Polypore (see p. 98). The thinking is that mycelium can be cultivated in substances that are cheap or unusable for other purposes, including refuse.

More information about edible mushrooms can be found in vol. 2.

shiitake

Mushrooms cultivated in a mine

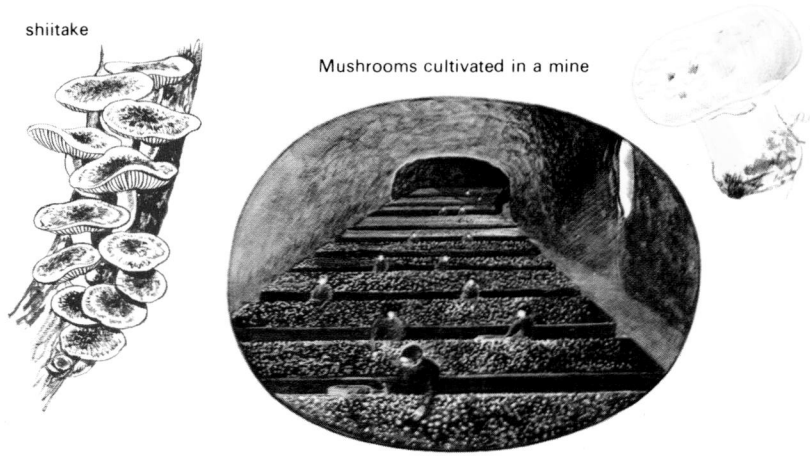

Poisonous fungi

Every year there are some cases of poisoning from eating certain fungi. Tales from many ages, and more recently the press, often mention mushroom poisoning. Unfortunately, lurid imaginations and a love of the mysterious have made them all too sensational. In recent years our ideas about the poisonous qualities of some fungi have had to be revised.

Among the many thousand types of large fleshy fungi, about ten are very poisonous. A large number are described as poisonous, inedible or unsafe, simply because nothing is known of the substances they contain. There have been countless examples of so-called 'safe' methods for telling whether a fungus is poisonous or not: if it turns silver black, if it clots milk, or if animals avoid it. None of these methods has any value. A fungus which is deadly poisonous to man may well be harmless to certain animals. There is only one sure way to tell whether a fungus is poisonous: eat it. Although modern chemical analysis can tell what substances a fungus contains, the effects of these substances on humans are uncertain. Experience, dearly paid for many times over, has taught us which fungi are edible and which are poisonous.

The most poisonous mushrooms are the Death Cap and the Destroying Angel. Several poisons have been isolated from them. The two most important, phalloidin and amanitin, are each composed of a number of substances, phallotoxins and amanitatoxins respectively. Phalloidin affects the tissues of the liver and destroys the cell membranes. However, there is some doubt about the exact role of phalloidin in poisoning cases. The amanitatoxins, on the other hand, are certainly deadly poisons. They destroy the nuclei of the liver cells and prevent protein synthesis. Also, the kidneys may be damaged. Both groups of poisons cause massive disorders of the digestive system (see p. 30). The symptoms do not appear for several hours or days. and often lead to death a few days later.

Among other very poisonous fungi, some of the major ones are certain species of *Inocybe*, for example *Inocybe patouillardii*, and *Cortinarius*, especially *Cortinarius speciosissimus*, Fly Agaric and Panther. The Fly Agaric (see vol. 2) contains muscarin and other poisons, although the muscarin is in much smaller quantities than in the *Inocybes*. Muscarin causes general poisoning, but the poison in *Cortinarius speciosissimus* – orellanin – has a special effect. It can fatally damage the liver and kidneys. Do not eat any species of *Cortinarius*. A number of other fungi are poisonous, but they normally cause only mild forms of poisoning with diarrhoea and vomiting. Many other *Cortinarius* species fall into this class, as well as species of *Paxillus*, *Clitocybe*, *Lactarius*, *Russula* and *Agaricus*. The bitter *Lactarius* and *Russula* species become completely harmless when cooked, as does *Gyromitra esculenta*, which is otherwise very poisonous and contains gyromitrin – another poison which damages the kidneys. The Common Ink Cap causes great discomfort if taken with alcohol. Its active component – coprin – acts in the same way as some anti-alcohol medicines such as Antabuse and Dipsan, but does not have their secondary effects.

In recent years there has been growing concern over poisons produced by moulds and other fungi which grow in and on foodstuffs and fodder. The diseases caused by these poisons are called mycotoxicoses, and the poisons usually go under the collective name of mycotoxins. Between two hundred and three hundred mycotoxins are known. Under certain moisture and temperature conditions, stored foodstuffs are easily attacked by poison-forming fungi such as *Aspergillus flavus* and species of the Glowing Mildew genus *Fusarium* (conidial fungi). *Aspergillus* forms, among other

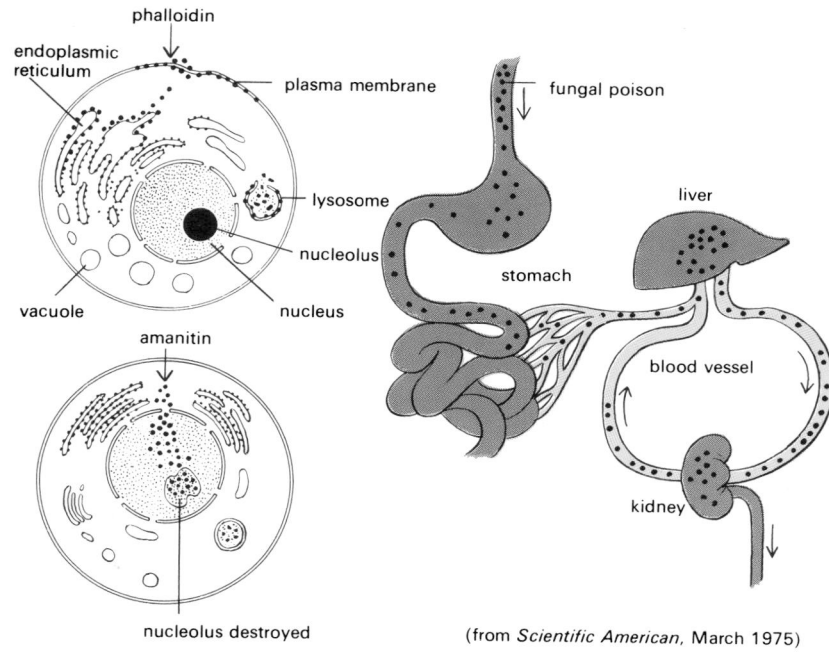

(from *Scientific American*, March 1975)

The illustrations show how the poisons from the Death Cap spread to the liver and kidneys, and how phalloidin and amanitin force their way into the cells of the tissue and affect or destroy their essential parts

things, the carcinogenic poison aflatoxin in badly-stored ground-nuts. Pig and chicken farmers face great problems with poisoning caused by fungal attacks on fodder. Both home-produced and imported foods are checked for the presence of such toxins. Although we should be aware of the dangers which the mycotoxins present, it should be pointed out that the poison's effect on humans is sometimes overstressed. One of the worst of the mycotoxins is alcohol!

In the last decade or two there has been scientific interest in the discovery of hallucinogenic fungi, and in research on them. Substances found in these fungi produce hallucinations. For thousands of years some types of fungi have had a significant impact on the Indian cultures in Mexico and elsewhere in Central America. The medicine-men ate fungi and put themselves into a trance, like the Siberian shamans who ate the Fly Agaric because they thought it was holy and a means of communication with the gods. These fungi were mainly species of the genera *Psilocybe*, *Panaeolus* and *Stropharia*. The substance psilocybin, isolated from these fungi, has effects which resemble those of mescalin and LSD. The drugs are being tested medically at present. They create changes in consciousness, and may well be of use in treating the mentally ill.

The stone mushrooms of Guatemala are between two thousand and three thousand years old and were probably religious symbols. Together with other ancient representations they indicate that there was some form of mushroom cult throughout the whole Maya culture. Some scholars maintain that there have been mushroom cults in other parts of the world.

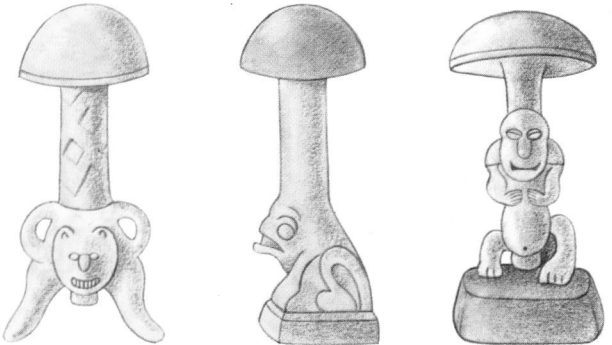

Stone mushrooms from Guatemala

Otidea onotica ❧

(Pers.) Fuckel (from Latin *otidium*: little ear and *onoticus*: little donkey's ear)

Hare's Ear

During the autumn months, where the soil is sandy, brownish or sometimes brilliant red fungi grow in pine-litter or leaf-mould or along newly-made roads. The majority of these disc fungi or Discomycetes are small and inconspicuous. Many grow on the dead remains of plants or on animal droppings. Some of these are shown on p. 43. The larger ones often grow in the soil, such as the genus *Otidea*, which has not been fully investigated, but includes perhaps some dozen species in Britain. Several are yellowish.

The Hare's Ear is cup-like, with one side deeply indented and the other side extended so that it looks like an ear. The flesh of the cup is thin, and can grow as long as 10 cm and as broad as 6 cm. The inside is orange-coloured with a tinge of salmon. The stalk is short and white at the base. This fungus is edible and has a pleasant smell.

Not all species of *Otidea* are ear-shaped. There is a brown, more regularly-shaped one, *Otidea cochleata* ❧ (L. ex St Amans) Fuckel (from Latin *cochleatus*: shell-like).

Hare's Ear
often grows in clusters
in pine-litter or
leaf-mould during
August and September

Aleuria aurantia ❧

(Fr.) Fuckel (from Greek *aleuron*: flour, and Latin *aurantiacus*: orange-coloured)

Orange Peel Fungus

The inside of this cup-shaped fungus is the brilliant colour of red lead. The outside is whitish and mealy, hence the Latin name for the genus. The cup is up to 10 cm broad, and is often wavy and irregular because it grows in clusters. Has a mild scent and is edible.

Many of the larger cup fungi are brownish. Another one, representing another family, is the dark brown cup fungus *Peziza badia* ❧ Pers. (from Latin *peziza*: stalkless fungus, and *badius*: chestnut-brown).

Orange Peel Fungus
grows in August and
September, often at the
side of newly-made
roads

Otidea onotica

Otidia cochleata

Aleuria aurantia

Peziza badia

Gyromitra esculenta ✦(☉)

(Pers.) Fr. (from Greek *gyros*: whorl and *mitra*: cap, plus Latin *esculentus*: edible)

These are spring fungi, and grow in sandy coniferous woods, especially where the soil has been disturbed, and often in places where trees have long since been felled. They are not uncommon in certain years and are fairly widespread, although they occur more frequently in northern parts of the country.

The cap is heavily embossed and wrinkled, and looks like a brain. The edge of the cap is often rolled inwards and embedded in the stalk. The cap grows to 10 cm in height and 15 cm in width. The colour varies from light brown to coffee or chocolate. The light-brown or greyish-blue stem is 3–6 cm high and 2–3 cm thick. The scent is pleasant. Fresh specimens are poisonous. The poison can be removed by boiling for ten minutes (using a water-to-fungus ratio of two to one) or by careful drying. Cases are known where *Gyromitra esculenta* have been eaten uncooked and have not caused poisoning. One possible explanation is that there may be several strains, not all equally poisonous.

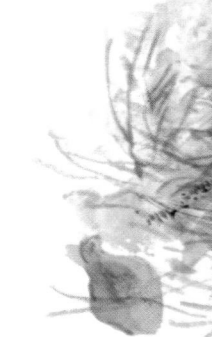

The brittle flesh of Gyromitra esculenta is *full of cavities*

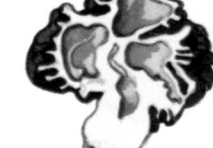

Neogyromitra gigas ✦(☉)

(Krombh.) Imai (from Greek *neos*: new and *gigas*: gigantic)

Bigger than the species described above. Specimens 25 cm wide have been seen. The cap is light brown with more regular and more vertically arranged creases. The cap rim is almost free. The stem is white and thicker. The fungus is almost without smell, grows under coniferous trees and is very rare. Its distribution is unknown. It is poisonous, but the poison can be removed by boiling.

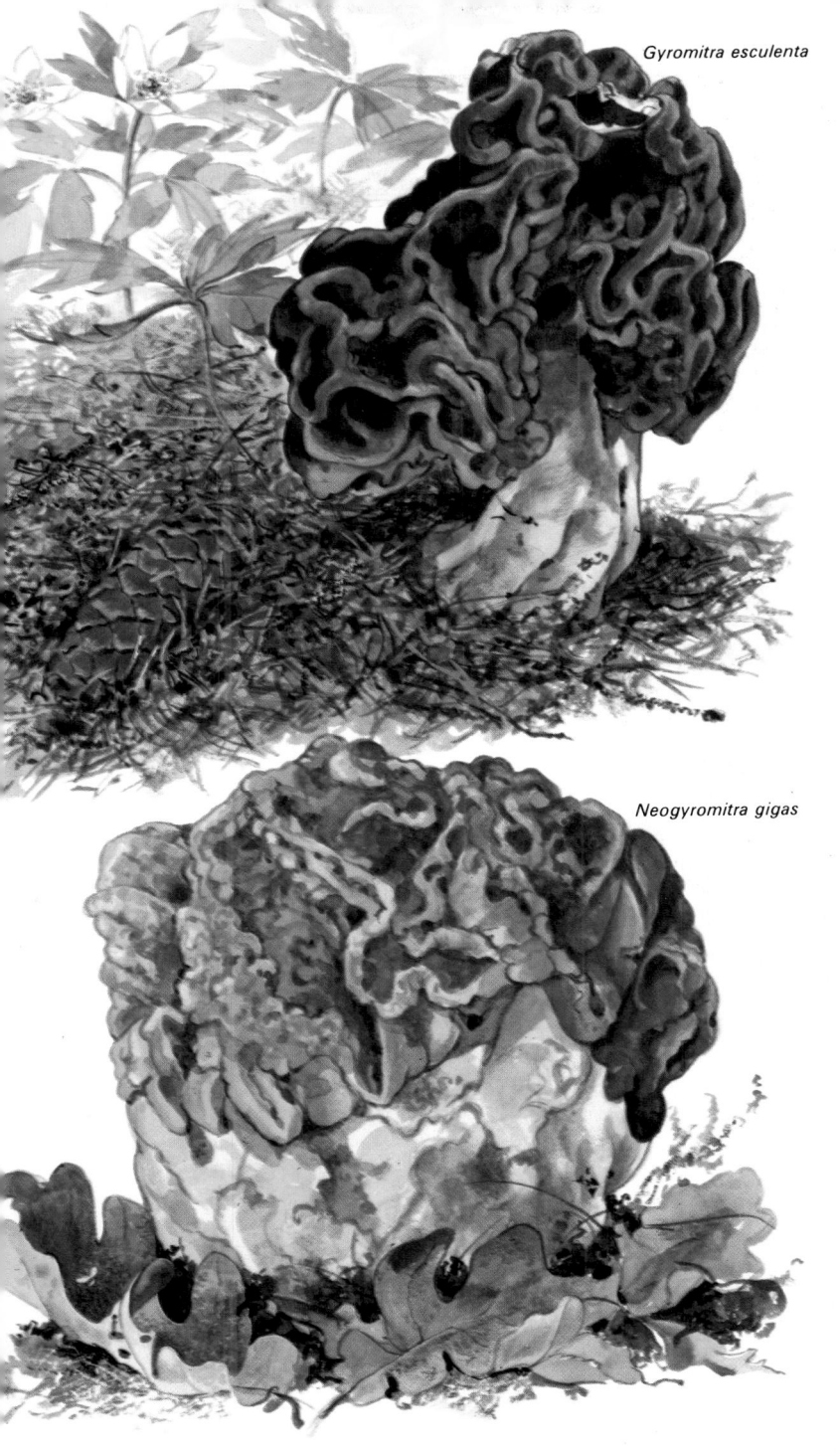

Gyromitra esculenta

Neogyromitra gigas

Gyromitra infula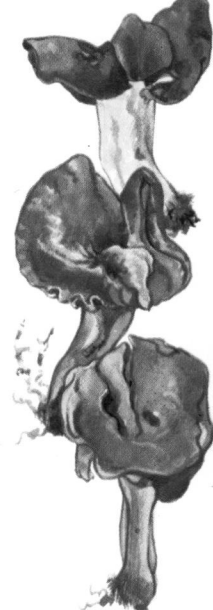

(Schaeff. ex Fr.) Quél. (from Latin *infula*: priest's hat)

In the autumn, fungi closely resembling *Gyromitra esculenta* are found on the rotting stumps and roots of pine trees and in the soil beside forest roads. These are *Gyromitra infula*.

It is more saddle-shaped than *Gyromitra esculenta* and not as wrinkled. The cap is very brittle. The cap grows to 12 cm in both height and width. The stem grows up to 12 cm high. The scent is faint. Fresh *Gyromitra infula* may be poisonous, so it should be boiled before eating. The flavour is not as delicate as *G. esculenta*. This fungus is rather rare in Britain, occurring mainly in the Scottish Highlands in October and November.

Helvella crispa ⊖

Fr. (from Latin *helvella*: the name of a cooking herb, later transferred to certain morels, and *crispus*: wrinkled)

Common White Helvella

White at first, and then acquires a yellowish tinge. It is smaller than *Gyromitra infula* but has long ridges and wrinkles on the stem. The cap is irregularly wrinkled, and can reach a height and breadth of 6 cm. The stem can grow to 10 cm and widens a little at the base. This fungus has a mild scent and is edible. It grows on grassy ground under beeches and oaks and is a common species, as is its close relative the Slate Grey Helvella *Helvella lacunosa* ⊖ Afz. ex Fr. (from Latin *lacunosus*: with hollows). Found in woods during October and November.

There are various strains of Gyromitra infula. *Some are reddish brown with a grey-white or light brown stem. Others are dark brown with a bluish stem*

Helvella lacunosa

Gyromitra infula

Helvella crispa

Morchella elata ⊖

Fr. (from Old High German *morchela*: little root, and Latin *elatus*: raised up)

Occasionally found in May and June on grassy ground that has been cultivated, often where there has been a fire. Although they tend to favour limestone, a soil that is rich in nourishment is necessary. The commonest forms of this species have a conical cap and a more or less regular network of ridges. The cap rim merges directly into the cylindrical stem. The cap grows to 10 cm high and 4 cm broad. The colour is olive grey or grey-brown. In dry weather the veins are often darker than the honeycombed surface. The white stem grows up to 8 cm high and 2 cm thick. The scent is often mild and pleasant. Edible without boiling.

Various sub-species have been illustrated and described. Many researchers now think that these all stem from a few principal types, among them *Morchella elata* Fr. and the yellow-brown *Morchella esculenta* ⊖ Pers. ex St Amans (from Latin *esculentus*: edible), which has a more irregular pattern of cavities and a rounder cap. This is widespread, especially in southern and eastern England. Morels, especially *Morchella esculenta*, are greatly valued in central Europe, where, in their dried form, they are an important food.

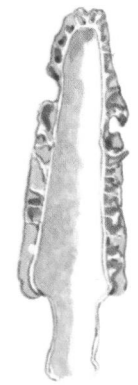

Morchella elata has a hollow cap and stem

Paxina acetabulum ⊖

(L. ex St Amans) O. Kuntze (after F. Pax, a German botanist, and from Latin *acetabulum*: vinegar bowl, little bowl)

Shaped like a cup or wine glass, has a short stem with long ridges which are often extended into branching veins on the underside of the cup. The cup grows to a breadth of up to 10 cm. The inside is dark brown, while the outside varies from off-white to pale clay. The whitish stem is 2–4 cm high. It has a mild scent and is edible. Grows in May and June in open woods and heathland in Britain, but is believed to prefer chalky soil. This fungus is normally smaller and darker in moorland country.

Morchella esculenta

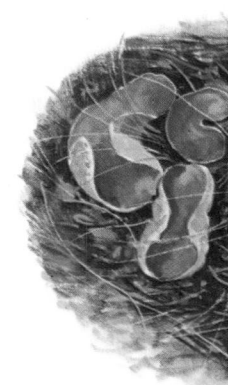

Morchella elata

Paxina acetabulum

Disciotis venosa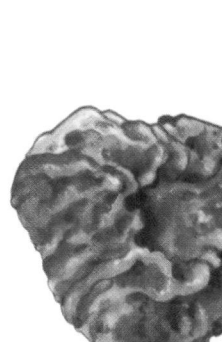

(Pers.) Boudier (from Greek *discos*: slice, plate, and Latin *venosus*: full of veins)

Looks a little like a flattened *Gyromitra esculenta*. It is found in shady places during April and May and on cultivated land and in leaf-mould.

At first it is bowl-shaped, then becomes almost flat, with wrinkles and grooves spreading from the centre towards the margin. It has a short, broad stem. The thick fleshy cup grows to a breadth of 15 cm. Its underside is whitish; its upper side is first yellowish brown, then darker brown. The stem grows to a height of 1 cm and is sometimes rather ribbed and often submerged in the soil. Has a striking smell (nitrous gas) and is not edible. A common species, growing on soil in deciduous woods in spring.

Discina perlata (Fr.) Fr. (from Latin *perlatus*: very broad) is a darker brown, with an ochre-coloured underside, and has no smell. It grows in the soil or on rotting stumps of conifers, and is edible. It sometimes grows in great quantity in Scandinavia where pines have been felled, but is extremely rare in Britain.

Hairy discomycetes

In damp soil and on rotten wood one often finds – in autumn especially – small Discomycetes that are hairy or downy on the outside. They often grow in groups. Two common kinds are illustrated.

Humaria hemisphaerica (Wiggers ex Fr.) Fuckel (from Latin *humus*: ground and *hemisphaericus*: hemispherical) has no stem and is rounded like a bowl, with a brown and hairy outside that feels like felt. The bowl is up to 2 cm wide, and the inside varies from a yellowy white to bluish grey. This fungus has a mild scent and grows on soil in woods between July and October.

Scutellinia scutellata (L. ex St Amans) Lamb. (from Latin *scutellatus*: having small shields) has the shape of a flattened cup, grows to a breadth of 1 cm and is a bright yellow-red. The layer of hairs on the outside is supplanted on the margin of the cup by black hairs up to a millimetre in length that look like eyelashes. It grows from spring to autumn, and is very widespread on wet ground.

Discina perlata

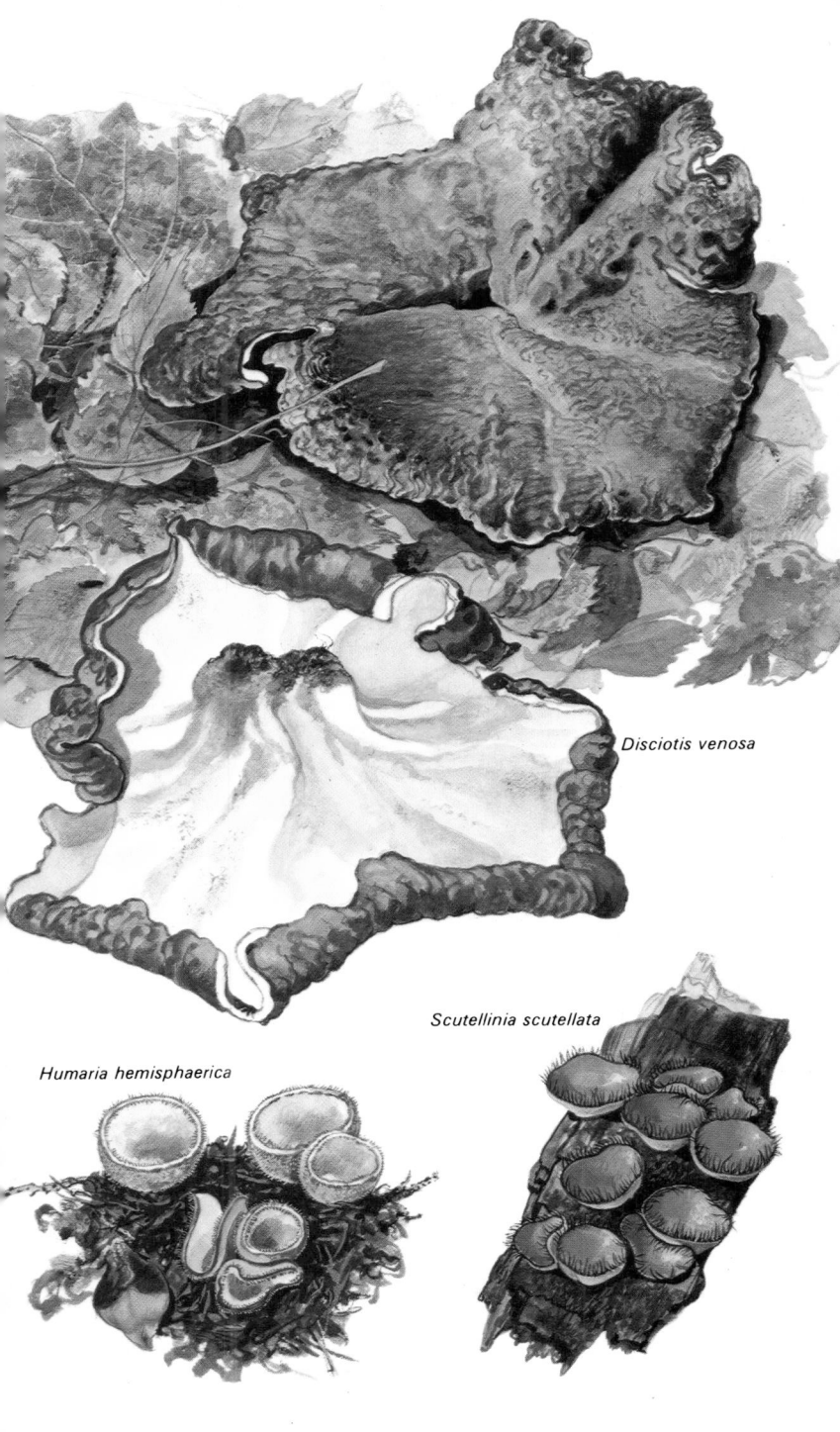

Disciotis venosa

Scutellinia scutellata

Humaria hemisphaerica

Discomycetes on wood ✣

During the entire growing season one finds on the dead and rotting parts of taller plants a host of tiny disc fungi, often only a few millimetres across. They belong to a group which contains many species, but when seen under the microscope are clearly different from the larger cup fungi and morels. Three species of the group are shown here.

Lachnellula subtilissima (Cooke) Dennis (from diminutive of *Lachnella* Latin *lachno*: woolly and *subtilissimus*: very fine) is up to 3 mm broad, and grows on the branches of fir trees.

Calycella citrina ([Hedw.] Fr.) Boud. (from Greek *calyx*: covering of a flower, plus Latin *citrinus*: lemon-yellow) is up to 5 mm across, and often grows in tight clusters on the decaying wood of deciduous trees.

Chlorociboria aeruginascens (Nyl.) Kan. (from Greek *chloros*: green, and *cibor*: drinking-vessel, plus Latin *aeruginosus*: verdigris green) grows on the decaying wood of deciduous trees and is up to 5 mm across.

Discomycetes are often shaped like tacks. The Lachnellula *species illustrated here has a fringed rim. The mycelium of* Chlorociboria aeruginascens *colours the wood a brilliant verdigris green*

Earth-tongues ✣

A group of morel-like fungi, growing mainly on the ground, belonging to the same group of discomycetes as the three species above. Some thirty species are known in Britain. Four of them are shown here.

Cudonia circinans (Pers.) Fr. (from Latin *cudo*: a helmet and *circinans*: growing in a circle) has an upper part which is head-like in shape. It is up to 7 cm high and occurs rarely in the coniferous woods of the Highlands from August to October.

Leotia lubrica Pers. (from Greek *leiotes*: smoothness, plus Latin *lubricus*: slippery) is the same size as the *Cudonia*. It is jelly-like and either sticky or slimy. It grows generally in damp deciduous woods from August to October.

Spathularia flavida Pers. ex Fr. (from Latin *spathula*: little spade and *flavidus*: yellowish) is shaped like a spade and grows to a height of 5 cm, generally in damp pine-litter from August to October. Rare.

Trichoglossum hirsutum (Pers. ex Fr.) Boudier (from Greek *tricho*: hairy, and *glosso*: tongue-like, plus Latin *hirsutus*: hairy) is shaped like a narrow club, often somewhat flattened, and grows to a height of 6 cm. It belongs to a group of brown and black species, and grows from August to November on ground rich in moss, especially sphagnum where animals have grazed.

The two groups of disc fungi discussed here represent a group which includes hundreds of species in Britain.

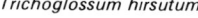

Trichoglossum hirsutum

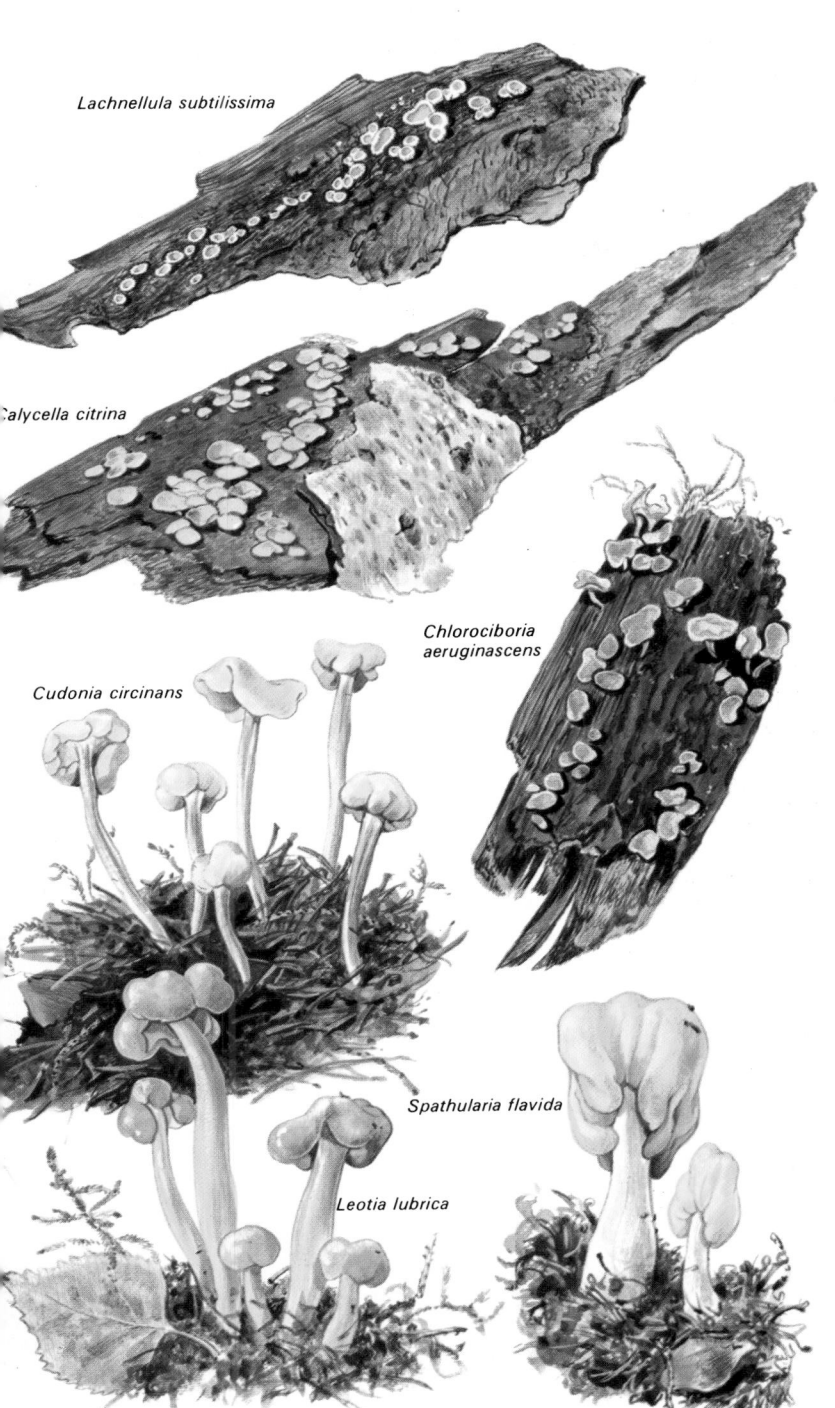

Lachnellula subtilissima

Calycella citrina

Chlorociboria
aeruginascens

Cudonia circinans

Spathularia flavida

Leotia lubrica

Choiromyces meandriformis ⊖

Vitt. (from Greek *choiros*: pig, and *mykes*: fungus, plus Latin *meandriformis*: creeper-like)

White Truffle

Truffles is a collective name for fungi that grow hidden in the soil and in some cases finally break through above the surface. They are generally spread by animals, which are attracted by the scent. The spores, hidden inside the fungus, often do not germinate until they have passed through the animal's intestines. The true truffles, including the prized black ones, are closely related to the morels (see p. 12). Many grow in southern countries and form fruitbodies during the winter.

Choiromyces meandriformis *grows underground, then breaks through above the soil*

Choiromyces meandriformis is irregular and knobbly, with a thin skin that is first greyish white, then yellow to brown. The inside is marbled. The size usually varies from that of a hazelnut to a walnut, but specimens sometimes grow as big as a fist. The scent is first weak, then strongly aromatic and finally disagreeable. This is a fungus highly valued as a spice. It grows on leaf-mould in deciduous and coniferous forests and also on clay soil in parks and gardens. It appears in August and September, and is rare except in certain local areas. This is the largest of the British truffles.

Many different fungus groups contain species of the truffle type. *Elaphomyces granulatus* ⊗ Fr. (from Greek *elaphos*: stag and *mykes*: fungus, plus Latin *granulatus*: fine-grained), for example, is an ascomycete belonging to a completely different group from the true truffles.

Elaphomyces granulatus *grows underground, with a hard skin around the dark mass of spores.* Cordyceps ophioglossoides *grows on it as a parasite*

Pyrenomycetes ⊗

There are many forms and species of Ascomycetes, but the largest group among them are the Pyrenomycetes, with more than a thousand species in Europe. They account for many of the small black 'dots' that one sees on dead vegetation. The spores are formed inside tiny fruitbodies, at most a millimetre broad, which sometimes cluster together to form a stroma (Latin: mattress, bed). Four kinds are shown here.

Cordyceps ophioglossoides (Fr.) Fr. (from Greek *kordyle*: swollen, and Latin *ophioglossoides*: like a snake's tongue). This has a club-like stroma.

Hypoxylon multiforme (Fr.) Fr. (from Greek *hypo*: under and *xylon*: wood, plus Latin *multiformis*: multiform). This is one of many species which form a cushion-like stroma.

Xylaria hypoxylon (L) Grev. (from Greek *xylon*: wood). Candle Snuff. This has a branching horn-like stroma. It is common on decaying tree stumps throughout the year.

Nectria cinnabarina (Tode ex Fr.) Fr. (from Latin *nector*: joined together and *cinnabarinus*: vermilion). Coral Spot. This is common on dead branches all the year round.

Hypoxylon multiforme *is common on dead birch branches.*

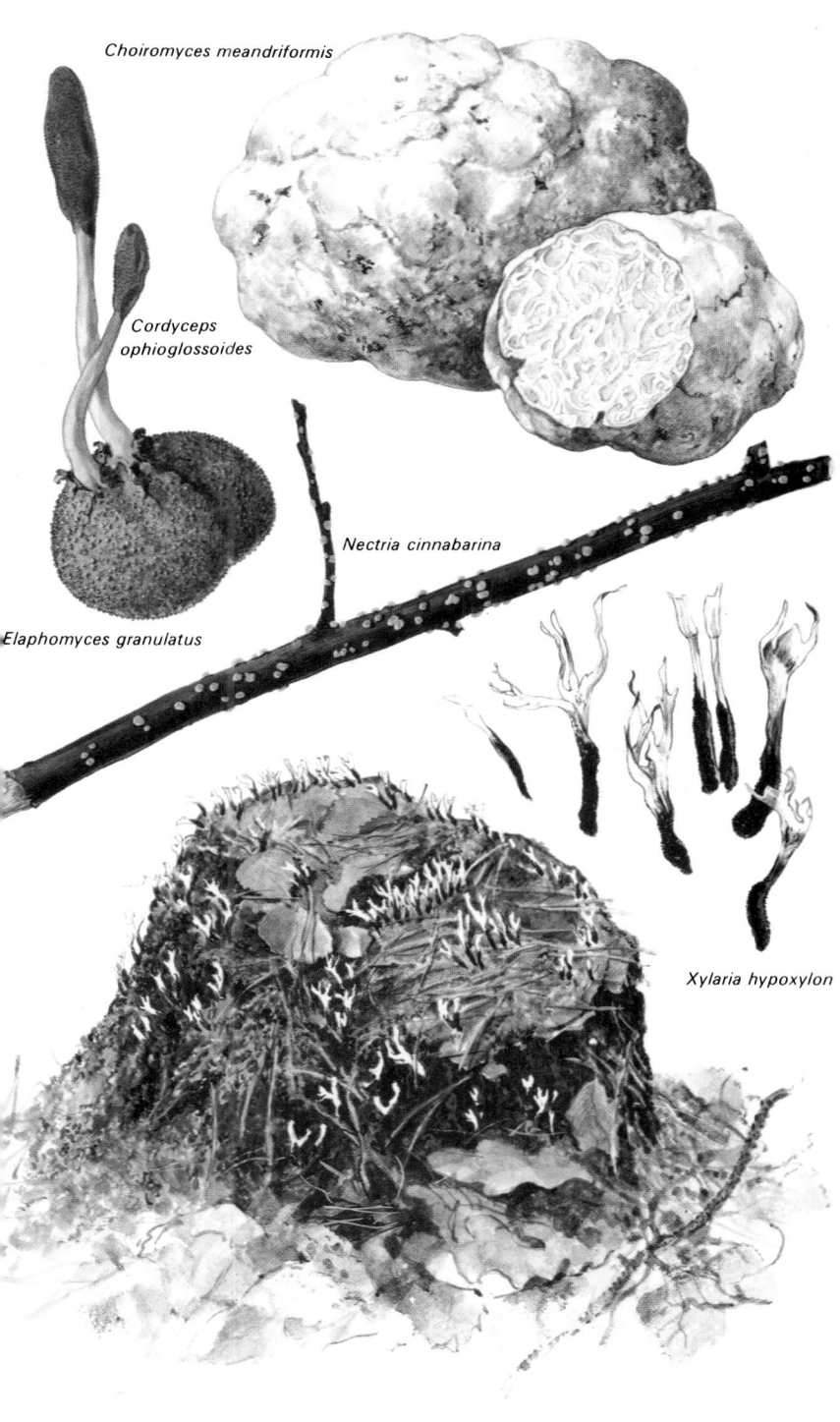

Choiromyces meandriformis

Cordyceps ophioglossoides

Elaphomyces granulatus

Nectria cinnabarina

Xylaria hypoxylon

Jelly fungi

Jelly fungi are found in damp weather on dead branches and trunks and on decayed stumps. They resemble morels, clavarias and hedgehog mushrooms in shape and colour. Because of their appearance when damp they are called jelly fungi. In dry weather they shrivel and are hard to see. As with agarics, boleti and other groups, their spores are formed on basidia, but the basidia of the jelly fungi are differently formed from those of, for instance, the agarics. There are about fifty species, of which six are shown here.

Pseudohydnum gelatinosum ⊖

(Scop. ex Fr.) Karsten (from Latin *pseudo*: false, *hydnum*: prickly fungus and *gelatinosus*: jelly-like)

This fungus is often tongue-shaped, up to 7 cm broad, and firm, but it is jelly-like and suggests a hedgehog mushroom with a short stem. The upper parts vary from light bluish grey to pale brown, often with small warts. The underside is somewhat lighter, and has awl-shaped barbs growing close together, 2–3 mm long. Edible but tasteless.

Jelly fungi often form irregular growths on decayed pine stumps from August to October

Çalocera viscosa ⊗

(Pers. ex Fr.) Fr. (from Greek *kalos*: pretty and *keras*: horn, plus Latin *viscosus*: sticky)

This is up to 10 cm high, forking repeatedly, and is yellow to orange. Unlike the true clavarias, it is tough and elastic. It grows commonly on rotten pine branches and stumps during September and October. *Calocera cornea* (Batsch. ex Fr.) Fr. is a similar unbranched species growing on frondose wood.

Many jelly fungi are, at the most, half a centimetre across, with a somewhat folded edge. They vary between a cushion and a shallow bowl-shape. Two yellow or orange-yellow species found on pine wood are *Guepiniopsis chryso-çoma* ⊗ (Bull ex St Amans) Brasf. (from Greek *chrysos*: gold and *coma*: hair) and *Dacrymyces stillatus* ⊗ Nees ex Fr. (from Greek *dakryon*: tear, *mykes*: fungus, plus Latin *stillatus*: drop-shaped).

Tremella foliacea ⊗ Pers ex Fr. (from Latin *tremo*: tremble and *foliaceus*: leaf-like) is one of the biggest of the jelly fungi. It is yellowish brown or cinnamon brown. At first it is wrinkled or brain-like, then lobate and thin-leafed. It grows up to 15 cm broad, mainly on deciduous wood.

Exidia glandulosa ⊗ Fr. (from Latin *glandulosus*: with glands) is almost black. It is cushion-shaped or crumpled and brain-like, and grows on dead oak branches.

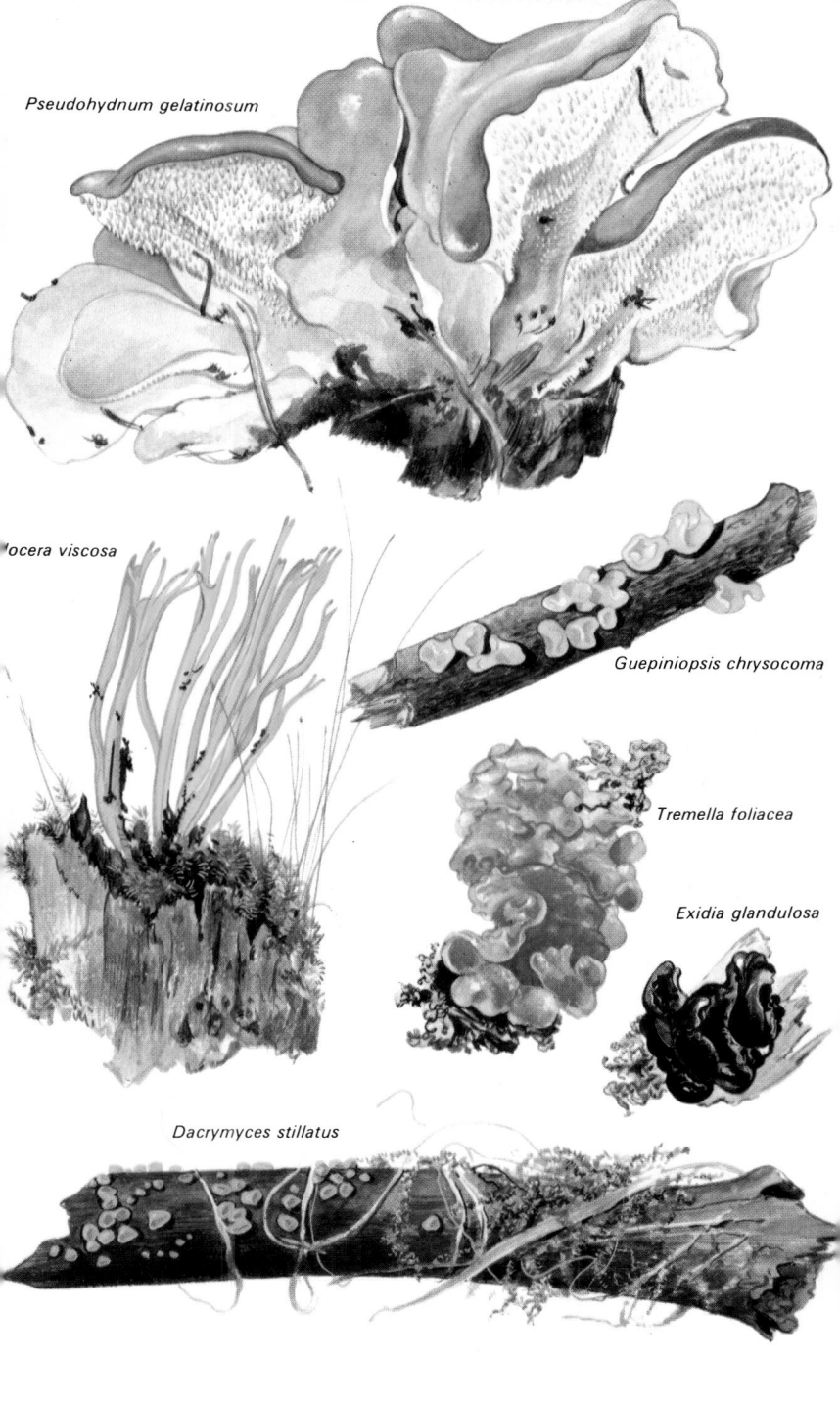

Pseudohydnum gelatinosum

ocera viscosa

Guepiniopsis chrysocoma

Tremella foliacea

Exidia glandulosa

Dacrymyces stillatus

Phallus impudicus ⊗

Pers. (from Greek *phallos*: phallus, and Latin *impudicus*: immodest)

Stinkhorn

In August and September an offensive smell, like that of rotting flesh, is sometimes noticeable in woodland. The Stinkhorn is usually the source. The Stinkhorn begins its life underground, shaped like a ball or an egg ('a witch's egg'), and has an elastic outer veil which soon bursts (see drawing below). The long white stem of the fully-grown Stinkhorn resembles a morel, and has a slimy olive-green substance (gleba) on top. In a cross-section of a 'witch's egg' we can distinguish the outer veil, a layer of jelly, the inner veil, gleba and the embryo of cap and stem (see drawing above right). Growth is very quick. The whole process often only takes one day and night, possibly two. The Stinkhorn's cap is bell-shaped, 2–4 cm high, with a network of raised veins, and covered at first with the gleba, which soon disappears. The stem is 2–3 cm thick, up to 20 cm high, and closely surrounded by the remains of the outer veil. This fungus is edible only in the egg stage. The gleba, which contains spores, attracts flies and other insects by its smell, and these then spread the fungus. The Stinkhorn belongs to a mainly tropical group of fungi, but some other species are found in the most southern parts of Britain. The Dog Stinkhorn *Mutinus caninus* ⊗ (Pers.) Fr. (from Latin *caninus*: dog) is about 10 cm high, with a white or faintly yellowish stem. When the dark green gleba disappears, the top is seen to be reddish. The Stinkhorn grows in deciduous woods, especially beech woods.

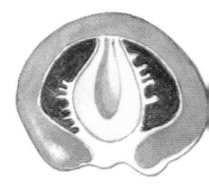

Cross-section of 'witch's egg'

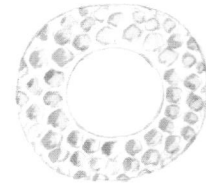

The Stinkhorn stem is porous and hollow

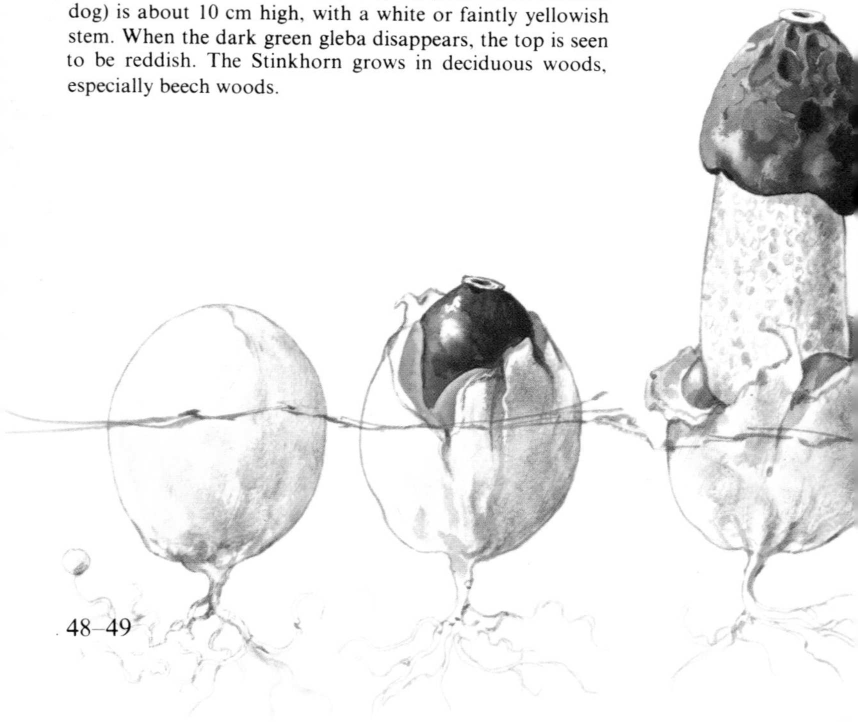

Lycoperdon perlatum ⊖

Pers. (from Greek *lykos*: wolf and *perdomai*: release wind, plus Latin *perlatus*: very broad)

Various kinds of puffball are found among fallen leaves and pine needles, in pastures and meadows, and on dry heathland. They are white and glistening in the autumn months, but their dry remnants remain through most of the year. The puffballs, including the genus *Lycoperdon*, which covers about a dozen species in Britain, belong to a large group of basidiomycetes – the gasteromycetes, or stomach fungi. The Stinkhorn also belongs to this group (see p. 48).

Lycoperdon perlatum is one of the commonest. It has a thick white outer veil with wart-like excrescences, and a thinner inner veil which surrounds the inner body of the fungus. The upper part of this forms the gleba (the spore-producing part). When fully grown the fungus is pear-shaped. It is greyish white with a tinge of yellow, and grows up to 7 cm high and 4 cm broad. In places where the excrescences have fallen off, which they do easily, it often shows a chequered pattern. It smells a little like a radish. Young puffballs with firm white flesh are edible. It grows in August and September, often in thinnish clusters in leaf or pine-litter, or on wood shavings or sawdust, but never on stumps.

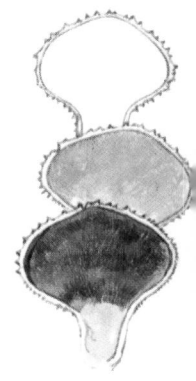

At first young puffballs have firm white flesh (top). Later the gleba becomes soft and yellow, then yellowish green

Lycoperdon pyriforme ⊖

Pers. (from Latin *pyriformis*: pear-shaped)

Stump Puffball

This fungus differs from the preceding species in having a smooth, though slightly granulose, outer wall. It grows in clusters on stumps and bits of decaying wood.

The Stump Puffball is pear-shaped, up to 5 cm tall and 3 cm broad. It is first whitish, then yellow-brown. The outer wall may, as in the illustration, break up into a pattern of angular fragments. The olive-coloured spores are puffed out through a little pore at the top of the inner wall. The scent is a trifle acrid, but young and firm specimens are edible, provided they are white inside. The Stump Puffball grows commonly throughout Britain during August and September.

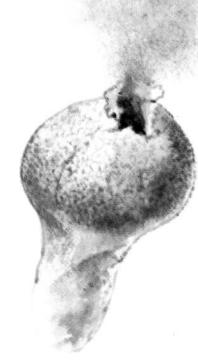

The spores escape into the wind through an opening in the top of the dry outer wall

Lycoperdon perlatum

Lycoperdon pyriforme

Calvatia excipuliformis ⊖

(Pers.) Perdeck (from Latin *calva*: cranium and *excipuliformis*: bowl-shaped)

The largest puffballs grow in meadows, pastures and heaths. One species, the Giant Puffball, *Calvatia gigantea* ⊖ (Pers.) Lloyd (from Latin *giganteus*: gigantic) is sometimes half a metre or more broad. Several other species may be 10–15 cm high and just as broad. These big puffballs have an outer veil which breaks into pieces when the fungus is full-grown and reveals the mature gleba. The genus *Bovista* ⊖ contains species which are rather like miniature Giant Puffballs.

    Calvatia excipuliformis is similar to *Lycoperdon perlatum*, but the outer wall is thin and grainy, or has small wart-like excrescences which soon disappear. When the thin inner veil bursts and comes away from the upper part of the fungus, and the gleba has dried and fallen off, the long dried stem often remains standing in winter. The fungus is greyish to brownish, up to 15 cm high and 4–7 cm broad, and often resembles a pestle. It may also have a yellowish tinge. The scent is mild, and fungi that are young, firm and white inside are good to eat. It grows from July to October and varies very greatly both in shape and colour. In highland country there is probably only a short-stemmed form.

The Giant Puffball appears occasionally on grassy land where there is plenty of humus. It always attracts attention when found

Scleroderma spp ⊗

(from Greek *skleros*: hard and *derma*: skin, plus abbreviation of Latin *species*)

Earth Balls

In woods and gardens with quantities of leaf-mould, but also in sandy soil, you may find stomach fungi with a leathery wall which resemble truffles. They grow from coarse, root-like threads of mycelium in the soil and are called Earth Balls. Unlike real truffles (see p. 44) they grow above ground. An example of an Earth Ball is illustrated. There are six species recorded from Britain but these are difficult to distinguish. The Earth Ball is like a potato and at first has a fleshy wall, which is usually yellowish brown and either grainy or warty. Later this becomes firm, and finally splits apart irregularly. The gleba is whitish at first, but soon becomes dark, with numerous lighter veins around the spore-bearing parts. At maturity the inner part disintegrates and the spores are spread by the wind. The Earth Ball smells acrid and is regarded as poisonous. The growing season is from August to October. Several species are rare, and their distribution is not fully known.

Calvatia excipuliformis

Scleroderma

Geastrum quadrifidum

Pers. ex Pers. (from Greek *ge*: earth and *aster*: star, plus Latin *quadrifidus*: four-lobed)

Earthstars

Earthstars, a very individual group of stomach fungi, are found not only under firs and pines, but also in deciduous woods and in sandy soil. There are about ten species in Britain. Some of them are locally common; others are rare or overlooked. Some are particularly fond of lime-rich soils.

Geastrum quadrifidum has an outer wall which, when the fungus is full-grown, forms four whitish lobes – occasionally three or five. The inner layer is grey or bluish grey, egg-shaped or pear-shaped, with a short but visible stem, and a clearly-marked pale band around a small smooth beak. The fungus may grow to 2–5 cm across and 4 cm in height. It grows in pine-litter. It is very easy to find earthstars in the winter, since individual fruitbodies may persist from September through to January.

Another earthstar, which also grows in pine-litter, is *Geastrum pectinatum* ✪ Pers. (from Latin *pectinatus*: comb-like). Fully-grown specimens have a dirty brown outer layer, 3–14 cm broad, with eight to ten lobes. The inner layer is 1–3 cm broad, and has an opening pore 3–7 mm long, grooved like a comb.

Earthstars grow underground at first and are almost roun. They have two layer an outer and an inne When the fungus gr. and matures, the ou layer splits into lobe These bend outwara the shape of a star a reveal the inner laye which encloses the gleba. The points of lobes turn downwara and the inner layer rises up on a star-lik pedestal. The spores are released via a be at the top of the inne layer

Crucibulum laeve ✪

(DC.) Kambley & Lee (from Latin *crucibulum*: crucible and *laevis*: smooth)

Bird's Nest fungi

Stomach fungi spread their spores in many different ways: by the wind, as with the puffballs; by insects, as with the Stinkhorn, and by large animals, as with the truffle types. One curious method is the manner in which Bird's Nest fungi, a group of species which grow on rotting wood, use waterdrops. This is described on p. 20.

Crucibulum laeve is first club-shaped, with a thick, yellowish brown wall, varying on the outside from hairy to smooth. Then it becomes cup-like, whitish on the inside, with lentil-shaped containers for the spores. It looks like a tiny nest with eggs in it. This fungus is common from June to October on frondose and coniferous twigs.

rucibulum laeve

*Geastrum
quadrifidum*

Geastrum pectinatum

Resupinate fungi 😵

A large number of fungi grow more or less pressed flat against decaying branches, trunks and stumps. Fungi growing in this way are resupinate (from Latin *resupinatus*: on its back). Many resupinate fungi from various groups of basidiomycetes play an important role in the natural cycle because they break down remnants of wood. Some cause 'storage rot'. The resupinates with a more or less smooth hymenium are the most important. These are divided into several families and include several hundred species. Three of them are illustrated.

Corticium laeve Fr. (from Latin *cortex*: bark and *laevis*: smooth) consists of dirty yellow or ochre spots, with an edge of fine white threads. The spots soon grow together into larger, extended fruitbodies, varying from smooth to granular. The species is common in coniferous and deciduous forests throughout most of the year. It is especially common on stored timber.

Stereum hirsutum (Willd. ex Fr.) S. F. Gray (from Greek *stereon*: hard, and Latin *hirsutus*: hairy) is not completely pressed against the substratum but has a protruding wavy margin. The hymenium is smooth or granular, first orange-yellow, then greyish yellow. Grows on deciduous wood and causes white rot on stored timber, mainly birch, beech and oak. It is extremely common.

Stereum rugosum (Pers. ex Fr.) Fr. (from Latin *rugosus*: wrinkled). This seldom has a protruding margin. The hymenium is yellowish to yellow-red, flecked with grey, and if damp it reddens when touched. This species grows on deciduous wood and also causes white rot.

Exobasidium vaccinii 😵

(Fuckel) Woronin (from Latin *ex*: out of, and *basidium*: a spore producing structure, plus *Vaccinium*: bilberry)

This belongs to a highly-specialized group of parasitic basidiomycetes which do not form any real fruitbody, but grow on the underside of the leaves of cowberry, whortleberry and related plants. The basidia grow directly from the mycelium. It distorts the leaves so that they become indented to form a spoon-like shape, varying in colour from white to red.

Exobasidium vaccinii

The upper side of Stereum hirsutum *has felt-like hairs*

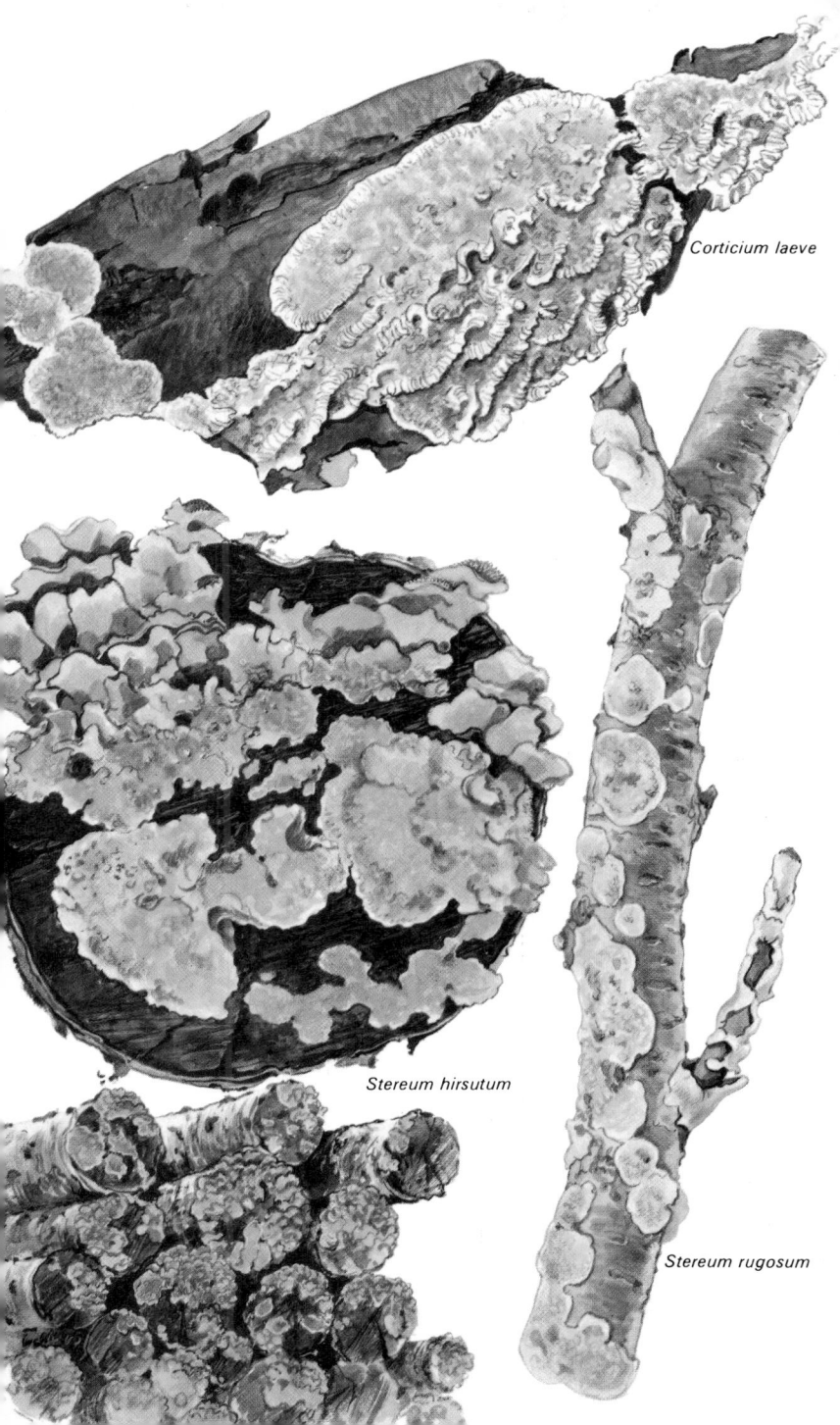

Corticium laeve

Stereum hirsutum

Stereum rugosum

Thelephora terrestris ⊗

Ehrh. ex Fr. (from Greek *thele*: wart and *phoros*: bearing, plus Latin *terrestris*: growing on ground)

In sandy places where the ground has been disturbed one often finds mussel-shaped, tough, grey-brown fungi with short stems. They may overlap like roof-tiles, and grow between August and December. The upper side is hairy and the underside granular. Species of this group are closely related to the resupinates, but there are some microscopic differences. *Thelephora terrestris* is one of the common species, 2–5 cm broad, light to dark brown, with a lighter margin that is often fringed. It often grows in the form of a rosette.

Thelephora palmata ⊗ Scop. ex Fr. (from Latin *palmatus*: like a hand) is a closely related, easily recognizable species. It is grey-brown, with branches like fingers, and has an unpleasant smell. It grows from September to November in the soil of pine woods, and is fairly common.

Thelephora palmata

Serpula lacrymans ⊗

(Schum. ex Fr.) S. F. Gray (from Latin *serpere*: to wind and *lacrimans*: weeping)

Dry Rot Fungus

Under favourable growth conditions, fungi can almost always destroy wood. There is a great risk of fungal damage in damp, badly ventilated wooden houses, especially from the ubiquitous Dry Rot Fungus. It grows in two ways: by mycelium inside the wood and by long strands of mycelium, 3–4 mm thick, outside the wood. These strands, which come from the mycelium inside the wood, can grow out across brickwork, stone and concrete, and attack timber several metres away from the site of the first attack. Moreover, the Dry Rot Fungus can attack dry timber. It can exude water, which can be seen coming from the fungus in drops (cf. the Latin name). The wood turns brown and brittle, and disintegrates or crumbles to dust. At first its fruitbody consists of thin fluffy mycelium. Later it is up to 1 cm thick, varies from dirty yellow-grey to rust-brown, with a labyrinth of wrinkles. The margin is whitish and swollen, and is often raised from the surface. The fungus can cover areas of a square metre, and grows mainly in houses.

Another fungus with a veined, wrinkled hymenium is the *Phlebia radiata* ⊗ Fr. (from Greek *phlebus*: vein, and Latin *radiatus*: radiating). This is pale flesh colour or orange-red, and grows on fallen branches in damp woodland – mainly deciduous – from August to December. It also causes white rot and damages stored timber.

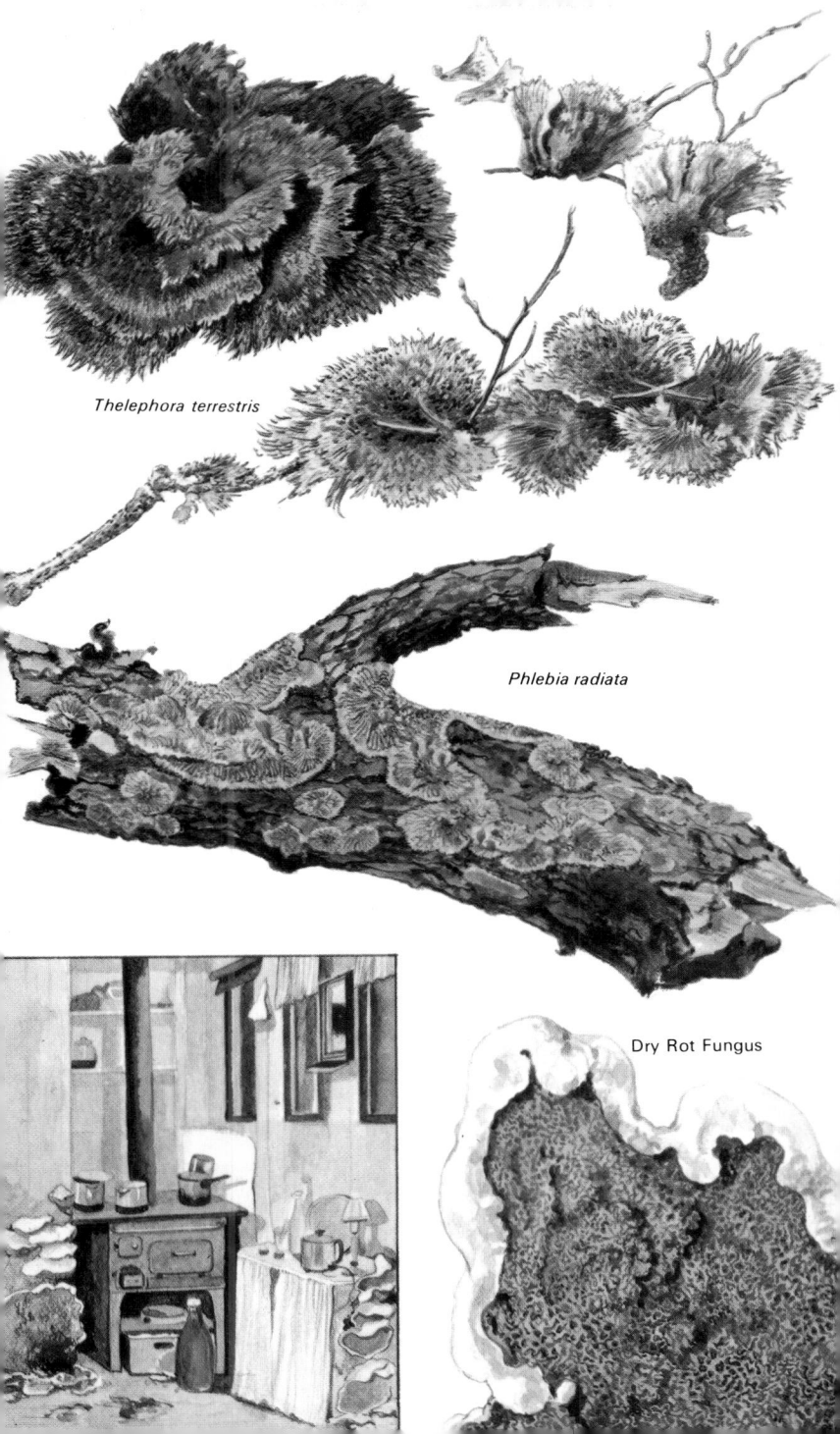

Thelephora terrestris

Phlebia radiata

Dry Rot Fungus

Club fungi

In autumn, among fallen pine needles, and among the fallen leaves of beech woods grow various sizes of yellowish club-like fungi. These are the club fungi or clavarias. They may represent species of the genus *Clavariadelphus*, the simplest form of these basidiomycetes. At one time all clavarias were regarded as belonging to one family. Now the inter-relationship of this very complex group has been investigated, and new groupings have been created which seem to resemble more closely the natural ones.

Clavariadelphus truncatus ⊖

(Quél.) Donk (from Latin *clava*: club, and Greek *adelphos*; brother, plus Latin *truncatus*: truncated)

A cylindrical or club-shaped fungus, 6–15 cm tall, with a seemingly severed top 2–5 cm broad. Its colour is yellow to ochre. Older specimens may be rust-coloured. The outside is more or less wrinkled, and is spore-bearing, while the top is sterile. Although the effect of the whole is solid, the inside has the expected fungoid texture. The scent is mild, and the taste is mild and sweet. The fungus is edible. It grows on decaying pine needles, and is common in August and September.

Another closely related species, equally large, is *Clavariadelphus pistillaris* ⊖ (L. ex Fr.) Donk (from Latin *pistillaria*: like a pestle), which has a more rounded top and a bitter taste. It seems to favour limestone and grows in deciduous woods, especially beech woods.

Clavariadelphus ligula ⊖

(Schaeff. ex Fr.) Donk (from Latin *ligula*: little tongue)

Very common under old pine trees, where the carpet of decaying vegetation has become sufficiently thick. It can grow by the thousand, and is therefore easy to spot.

It grows at the most to 10 cm in height and 1.5 cm in breadth. It starts off yellowish white, then becomes red-yellow or brownish, and is often flattened out like a tongue. The scent and the taste are mild. The fungus is edible.

Yet another species of club fungus is the easily recognizable, very rigid *Clavariadelphus fistulosus* ⊖ (Fr.) Corner (from Latin *fistulosus*: equipped with a pipe). It is brownish and grows to 20 cm high, but only 0.5 cm wide, on rotting wood, and is without taste or smell.

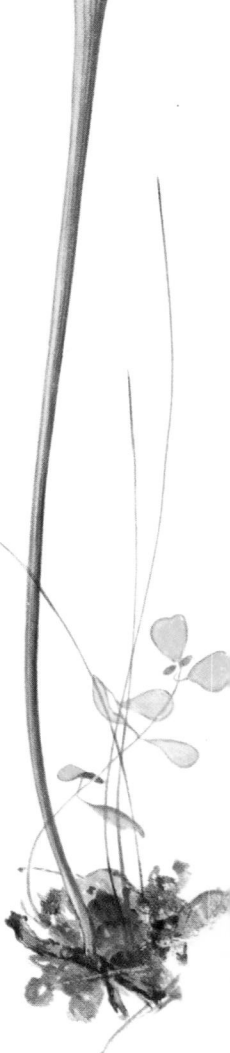

Clavariadelphus fistulosus

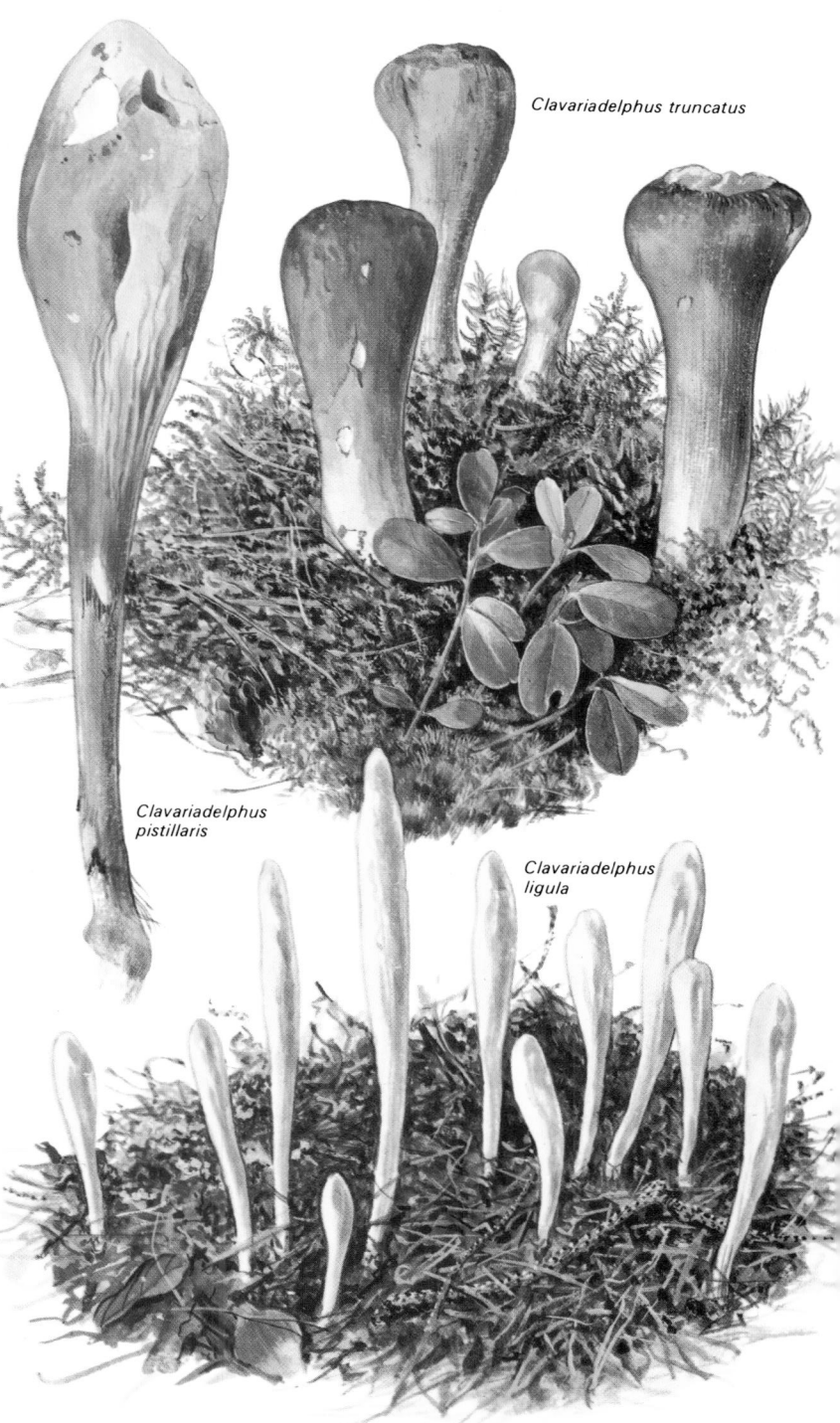

Clavariadelphus truncatus

Clavariadelphus pistillaris

Clavariadelphus ligula

Clavarias

The yellowish, fleshy and branching clavarias are very familiar to many mushroom-pickers. From the scientific point of view, however, the different species are extremely hard to separate. In this section we illustrate and describe some varieties that could be called yellow clavarias. We start with a white species of the genus *Clavulina* (from Latin *clavulina*: little club) which varies a good deal in shape and grows among leaves and pine needles, mainly in wooded country. Another genus whose species are predominantly club-like or only slightly branched is *Clavulinopsis* (from Latin *Clavulinopsis*: like *Clavulina*). An example of an unbranched species belonging to this genus is shown in the bottom right-hand corner of the opposite page. Species of *Clavulinopsis* can be white, yellow, orange or red. There are also some branching species within this genus, which are characterized by a waxy surface. One of these is shown at the bottom of this page. In Britain there are probably around thirty species of *Clavulina* and *Clavulinopsis*.

Clavulina cristata ⊖

(Fr.) Schroeter (from Latin *cristatus*: crested)

This common and widespread fungus is whitish, sometimes with a tinge of yellow or grey, and normally has a distinct stem and comb-like branches. It grows to 7 cm in height. The scent is pleasant and it is one of the few palatable clavarias.

Ramaria invalii ⊗

(Cott. & Wakef.) Donk (from Latin *ramosus*: branching, and *Inval*: place near Haslemere, Surrey)

The genus *Ramaria* contains many brightly-coloured yellow species. A few are whitish. They branch profusely in all directions, and are often fleshy. The *Ramaria* species grow mainly among decaying leaves and pine needles. There are some hundred species altogether, of which perhaps forty are to be found in northern Europe. Several are suspected of being poisonous. The crowded branches, which grow from a barely perceptible stem, are brownish yellow, slim, pointed and divide repeatedly. The fungus grows up to 8 cm. The flesh is whitish and tough and tastes bitter. Clavarias which are not edible often grow under fir trees from August to October. Several closely-related species turn bluish or blue-green when the flesh is broken.

Clavulina cristata

Clavulinopsis sp.

Ramaria invalii

Yellow Ramarias

Ramaria sp. (from Latin *sp.*: abbreviation of *species*)

In both deciduous and coniferous woods there are large, fleshy, yellowish clavarias. When young they are often sunk in moss or hidden among leaves and therefore hard to find, but when full-grown they are very conspicuous. Often several can be found together.

The yellow Ramarias vary in shape. Most books on fungi give the yellow Ramaria the Latin name *Ramaria flava* ⊖ (Schaeff. ex Fr.) Quél. (from Latin *flavus*: light yellow) but it is not clear to which of the yellow species this name should apply. Our source for the Latin names of many fungi is *Systema Mycologicum* (1821–8) by the Swedish mycologist Elias Fries. Fries based his names on various eighteenth-century works, one of them by Jakob Schaeffer, who did his research in southern Germany. Fries studied the fungi of Småland and the Uppsala region of Sweden, and tried to interpret the mid-European species through his own discoveries. In *The Edible and Poisonous Fungi of Sweden* (1861) he gives *Ramaria flava* the Swedish name 'The High Yellow Clavaria'. Fries describes it as follows:

Ramaria formosa

The High Yellow Clavaria forms a tight clump of upright, round, lemon-coloured or sulphur-coloured branches. They grow close together, are of almost the same height, and join together below in a thick whitish trunk. The branches are extremely brittle and break off easily, and the white seed-dust which this produces is the easiest way to distinguish it from the Ochre Clavaria, which is darker and has yellow seed-dust. It is the latter (*Clavaria aurea* Schaeff.) which Persoon and most researchers have understood to be *Clavaria flava*. The High Yellow Clavaria occurs throughout the Kingdom in deciduous woods; sometimes in pinewoods also.

The species shown at the top of the facing page may be *Ramaria flava*. The lower one with the deeper colour may be *Ramaria aurea* ⊖ (Fr.) Quél. (from Latin *aureus*: golden yellow).

Many of the yellow Ramarias are mildly poisonous. One of these is a yellow-red species *Ramaria formosa* ⊠ (Fr.) Quél. (from Latin *formosus*: pretty). This grows in beech woods. It has pink branches with lemon-coloured tips, and the flesh of both the stem and the branches is white. Those which are edible have pale gold flesh in their branches. The difficulty of telling the yellow species apart means that it is best to avoid eating them. Finally, another difficulty in sorting out the yellow branching species is that there are often dwarf forms which are less clearly branched. A dwarf form of this kind is shown bottom left on the opposite page.

High Yellow
Clavaria

Ramaria sp. (dwarf form)

Ochre Clavaria

Ramaria botrytis ☻

(Fr.) Ricken (from Latin *botrytis*: like a bunch of grapes)

'In the beechwoods of southern Sweden this beautiful species often grows *en masse* in rainy summers. In very dry summers I have usually sought for it in vain. In central Sweden it is rare, but it has been found at Uppsala and in Södermanland.' Thus wrote Elias Fries in 1861. It is uncommon in Britain. Young specimens are easily recognized, with their white to pale yellow branches – fleshy and tipped with bright red. On the other hand older specimens, yellowish and without the red tips, may be hard to distinguish from other *Ramaria* species. It is one of the few of its genus which are good to eat. The tips, however, can be bitter and should be cut away. This fungus has a more robust stem and thicker branches than the yellow clavarias. It grows to 15 cm in diameter and occurs during August and September. Its scent is pleasant, and its flavour is essentially mild.

Sparassis crispa ☻

(Wulf. ex Fr.) Fr. (from Greek *sparaxis*: raggedness, and Latin *crispus*: curly)

With numerous flat, ribbon-like branches, curling like waves and sometimes growing together, this hardly resembles any other fungus but is probably related to the clavarias. They are conspicuous because they are large – often the size of a cabbage head. The genus *Sparassis* is widely distributed but this species is only common in localized areas. The fungus grows at the base of conifer stumps. It is more common in the damper areas of central Europe. Two very rare species also occur in Britain. *S. laminosa* is said to grow on oak, and *S. simplex* occurs in pine-litter. Some other species grow in some areas of rainy, virgin pine forests on the west coast of North America.

Sparassis crispa grows to 40 cm wide and 30 cm high. The branches at its edge are whitish yellow to brown-yellow, curly and 1–2 cm broad. The stem, which grows to 8 cm high and 4 cm thick, often looks like a root. The scent is aromatic, and the flavour of the young fungus is exquisite. It grows during August and September.

The stem of Sparassis crispa *often looks like a root*

Ramaria botrytis

Sparassis crispa

Gomphus clavatus ☉

S. F. Gray (from Greek *gomphos*: plug, and Latin *clavatus*: club-like)

This species may be found in mossy pine woods, and sometimes in beech woods. When young it suggests a large club fungus, and is regarded as a relative of that group. It probably flourishes on limestone or chalky clay soil. The cap is yellow-grey, up to 10 cm broad, unevenly curved and often lopsided. The top is smooth, while the red-violet underside is venose. The stem is at first pale violet to red-violet, then paler and it merges directly into the cap. It grows to 10 cm high. The scent is weak and the taste mild. *Gomphus clavatus* is edible and grows in August and September.

Craterellus cornucopioides ☉

Pers. ex Fr. (from Greek *krater*: beaker, and Latin *cornucopioides*: like a horn of plenty)

Horn of Plenty

Because of its brownish to bluish colour, and more or less smooth hymenium, it is easy to distinguish the Horn of Plenty from the chanterelles, to which it is closely related. It grows both in deciduous and in coniferous woods. It is common in hazel groves and beech woods, where it is often hidden among leaves and in the shadows of trees, but it occurs also in mossy pine woods. The cap is funnel-shaped, 5–8 cm broad, and often has a somewhat curled edge. The stem grows to 12 cm high, and is a direct extension of the cap. The flesh is thin, and the fungus has a pleasant fruity smell and a delicious taste. It grows in August and September.

Another fungus which has sometimes been assigned to *Craterellus*, sometimes to the chanterelles, is *Cantharellus lutescens* ☉ Fr. (from Latin *lutescens*: turning gold). Its upper side is very similar to the Horn of Plenty, but more spread out and with small scales. The underside, however, is a clear orange-yellow. The ridges and veins on the underside of the cap are not so clearly marked as on the common Chanterelle, which explains its disputed classification. The scent is pleasant, and the fungus is edible and much sought after. It grows in damp places in mossy pine woods, often on the edge of swamps, during August and September.

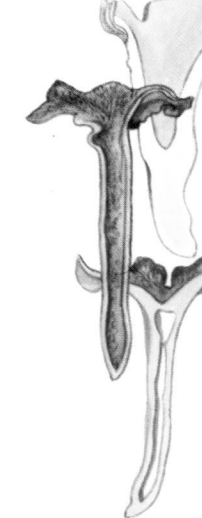

Gomphus clavatus *is fleshy with a solid stem. The Horn of Plenty and* Cantharellus lutescens *are slim with hollow stems*

When damp, the Horn of Plenty is almost black with a tinge of blue

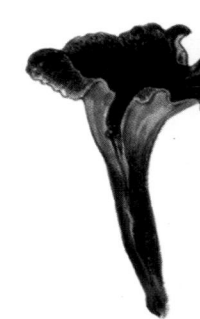

Gomphus clavatus

Craterellus cornucopioides

Cantharellus lutescens

Cantharellus cibarius ⊖

Fr. (from Latin *cantharus*: beaker and *cibarius*: table)

The Chanterelle

Found among beeches and oaks, in birch groves and at the edge of swamps where there are birch trees, in mossy fir woods and on pine-covered moorlands. The Chanterelle is particularly fond of trodden ground where it grows beside paths and roads. However, it must have living trees nearby because it is a mycorrhizal fungus (see pp. 18–19).

There are several forms of Chanterelle. A pale and fleshy form found in deciduous woodland is now regarded as a separate species (see below).

The Chanterelle is egg yellow, with no clear division between cap and stem. The underside of the cap has many fork-like veins and ridges, sometimes so pronounced that it resembles the gills of an agaric. The cap is at first uneven, then funnel-shaped with a curly edge. It grows up to 10 cm wide. The stem, 5–10 cm high, is the same colour as the cap. The fungus becomes flecked with brown when touched. The scent suggests dried apricots, and when eaten raw has rather a sharp taste. Properly prepared, it is one of the most delicious fungi. The Chanterelle is found between July and November but it grows slowly and needs a lot of rain.

False Chanterelle

Cantharellus pallens ⊖

Pilát (from Latin *pallens*: pale)

Pale Chanterelle

The Pale Chanterelle often grows in hazel groves

This chanterelle is more solidly built than the common one. It usually occurs earlier, in wet summers at the beginning of July, but has not yet been reported in Britain. The upper side of the cap is whitish to pale yellow, often with a touch of pink. The underside is egg yellow in older specimens. It grows to 15 cm in breadth and height and has the same excellent flavour as the Chanterelle.

The true chanterelles are sometimes confused with the tasteless but harmless False Chanterelle *Hygrophoropsis aurantiaca* (⊖) (Fr.) Maire (from Greek *hygrophoropsis*: like a wax agaric [*Hygrophorus*], and Latin *aurantiacus*: orange-coloured). This has a pale yellow to fire-yellow soft and downy cap, and many fork-like gills. It is a cantharelloid fungus – i.e. it resembles the chanterelles, but is not related in any way to that group. It is found in coniferous woods and heathland from August to October.

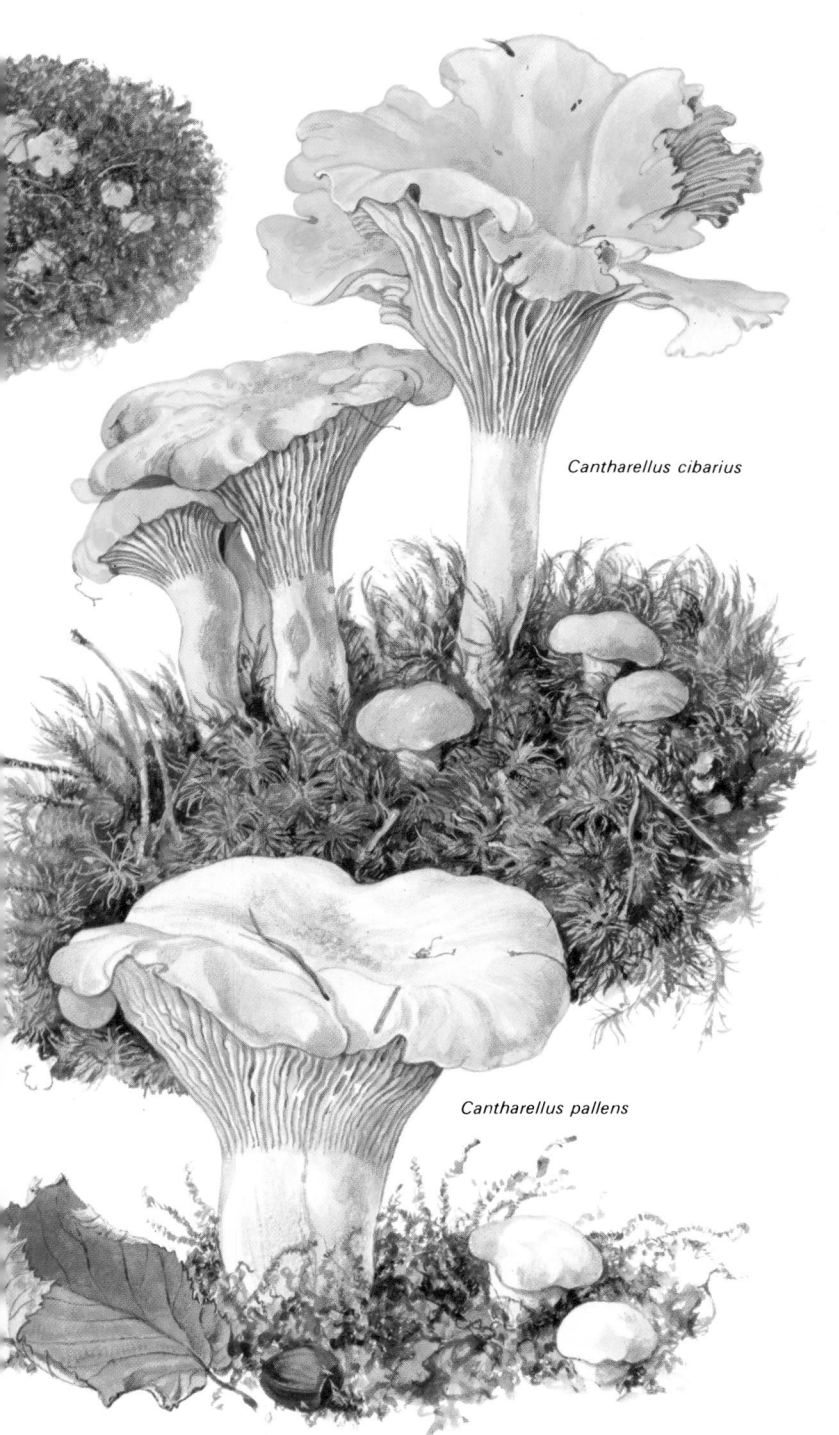

Cantharellus cibarius

Cantharellus pallens

Cantharellus tubaeformis ⊖

Bull. ex Fr. (from Latin *tubaeformis*: trumpet-like)

An autumn fungus, sometimes called the Autumn Chanterelle, grows in coniferous and deciduous woodland and in
certain years occurs in great numbers. The grey-brown caps
resemble withered leaves, and are therefore hard to find, but
once one is located more should be easy to find as it grows in
large colonies. The mycelium may be spread over areas
covering hundreds of square metres. The fungus can be found
both in chalky and in fairly sour soil. It is edible although the
flavour is a little inferior to that of the Chanterelle. It can still
be picked when other mushrooms have been killed by frost,
and during a mild autumn can be found as late as November.
This is sometimes confused with *Cantharellus lutescens* ⊖ ,
which probably thrives in chalky soil.

The cap flesh is thin. At first it is shallowly embossed, with
a rolled-back margin. Later it is deeply embossed, wrinkled
and scaly, with a curly margin. It grows to 6 cm in width.
The stem is 3–7 cm high, off-yellow to yellowish grey. It is
somewhat flattened in older fungi, and is hollow through to
the base. The scent is pleasant and the flavour mild. It grows
from September to November.

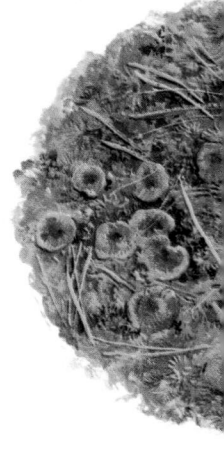

Cantharellula umbonata ⊖

(Fr.) Singer (from Latin *umbonatus*: with a hump)

This uncommon species is found in mountainous pine woods
among moss and lichen. It is an ash-coloured, relatively small
fungus, with many pale grey fork-shaped gills. Formerly
regarded as a chanterelle, it is now known to be a true agaric,
and yet another example of a cantharelloid fungus (see p. 70).
The cap grows to 4 cm broad and the gills acquire streaks or
flecks of pale red. The flesh is thin, with an in-turned edge.
The middle of the cap is somewhat sunken and has a distinct
hump. The stem is greyish and elastic, and grows to 5 cm. In
older specimens it is hollow. The fungus is mild, without scent,
and is edible. It grows singly or in small groups from August
to November.

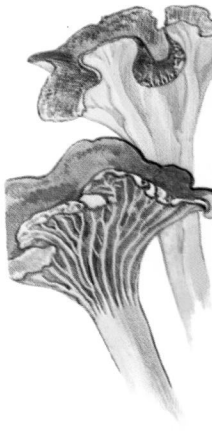

The Cantharellus
tubaeformis *cap has
blunt, repeatedly
forking ridges on the
underside. These are
yellowish to grey and
run down the stem.
The cap of*
Cantharellus lutescens
*has an orange-yellow
underside, with no
noticeable veins*

Cantharellus tubaeformis

Cantharellula umbonata

Hydnum rufescens ⊖

(Pers. ex Fr.) Fr. (from Greek *hydnon*: truffle [later arbitrarily transferred to certain spine fungi] and Latin *rufescens*: reddening)

Fungi environments differ greatly. Some species grow only in very special places and others can grow in both deciduous and coniferous woods, and in the most diverse climates. Species of *Hydnum* occur widely. *Hydnum repandum* grows especially plentifully in beech woods but it is also found in quantity in mossy coniferous woods and other deciduous woodland. *Hydnum rufescens* usually grows in damper places than *H. repandum*. Both represent a group of fungi with a spiny hymenium. All such fungi were once classed together as one genus, *Hydnum*. It is now known that the spine fungi are not a homogeneous group, but should be segregated into several genera belonging to different families. The nearest relatives of *Hydnum rufescens* and *H. repandum* appear to be the chanterelles. The cap of *H. rufescens* is ochre to orange-brown, up to 7 cm broad, and has dense, yellow-orange to salmon-coloured spines. The stem is 2–7 cm tall and up to 1.5 cm thick. This fungus has a pleasant smell and is edible. It is common between August and October.

Hydnum repandum ⊖

(L. ex Fr.) Fr. (from Latin *repandus*: turned upwards)

Bigger and fleshier than *H. rufescens* and lighter in colour. Often grows in pairs or clusters and thus becomes irregular in shape. The flesh sometimes turns rust-coloured when touched. The cap is cream-coloured to yellow-red, smooth, shiny in dry weather and up to 15 cm wide. The spines are whitish at first, then become the same colour as the cap and are brittle and of uneven length. The stem is whitish, 2–7 cm tall, distended at the base and often lopsided. Has a pleasant smell and is good to eat but like the chanterelles, it has a rather sharp taste when raw. The sharpness disappears with cooking, however, and this is one of the best edible mushrooms of late autumn, from August to November.

Hydnum rufescens

Hydnum repandum

Sarcodon imbricatus ⊗

(L. ex Fr.) Karsten (from Greek *sarxos*: flesh, and Latin *imbricatus*: with scales like roof-tiles)

These large spine fungi grow in coniferous woods, mainly in September. They are grey-brown, vary from slightly arched to funnel-shaped, and have a more or less scaly cap. Most commonly they have distinct dark scales and a paler base. They grow mainly in mossy places in coniferous woodland. Among the brownish fleshy spine fungi, which are now placed in a separate genus, there are some very unusual species. Some have a cap which is smooth at first, but then turns scaly. In some species the flesh turns bluish green when cut. Several of the unusual species flourish in chalky ground. The taste is often bitter or acrid, and none of these species are good to eat.

The cap of *Sarcodon imbricatus* is pale grey and 20–30 cm broad. It is rounded at first, then becomes somewhat funnel-shaped, with rough scales that are grey-brown to sepia. The spines are grey, 5–10 mm long, and crowded. The stem is pale grey, up to 8 cm tall and 3 cm thick. The scent is weak, and the taste is slight at first, but later acrid or bitter. If it is mixed with other mushrooms, its flavour takes over completely and can spoil an otherwise tasty dish. Grows from August to October.

Sarcodon imbricatus is mentioned in Swedish writings even before Linnaeus. We know from Celsius (1732) that it could be found in the Uppsala area. Linnaeus also mentions it in *Flora lapponica* (1737). He probably refers to specimens seen along the coast of northern Sweden.

Sarcodon imbricatus

Hericium clathroides ⊖

(Pall. ex Fr.) Pers. (from Latin *hericius*: hedgehog and *clathri*: lattice)

Ear-pick fungus

Some species of spine fungi grow on wood. They are found on stumps and rotten wood, and in exceptional cases on living trees. The prettiest, and the likeliest to attract attention, is *Hericium clathroides* which grows on fallen branches, and sometimes on living deciduous trees. It is rare but may be found on birch, aspen, beech and oak.

Hericium clathroides branches into a lattice pattern, up to 15 mm long, and has many densely-packed spines. It is at first white, then yellowish. The fungus can spread itself over large areas of the tree trunk. The scent is sour and the flavour radish-like. It grows from August to October.

Creolophus cirrhatus ⊙

(Pers. ex Fr.) Karsten (from Greek *kreas*: flesh and *lophos*: wisp of hair, plus Latin *cirratus*: curly)

This fungus also grows on wood, and is very irregular in shape. Usually it consists of mussel-like caps growing together, and suggests a soft polypore. When dry it is almost white with a pale tinge of yellowish red. When damp it is pink. The upper part of the cap has warts or tiny spines, and the underside has loose-packed spines, pale to yellowy red. It can be up to 10 cm thick at the point of its foothold. The fungus grows up to 10–15 cm long and 8–10 cm broad. The scent is pleasant, and young specimens have a good flavour. It grows on the stumps of deciduous trees, mainly birch, from August to October. It is not common, but more frequent than *Hericium clathroides*.

Auriscalpium vulgare ⊖

S. F. Gray (from Latin *auris*: ear and *vulgaris*: ordinary)

Ear-pick Fungus

This small spine fungus, with its dark felt-like kidney-shaped cap and its lateral stem, is common but difficult to see and is therefore easily overlooked. The cap is 1–2 cm wide and the stem up to 6 cm high. Grows mostly on pine cones but single occurrences on fir cones are also known. Grows only at times of year when there is no frost, especially during September and October.

Hericium clathroides

Creolophus cirrhatus

Hydnellum aurantiacum ⊗

(Batsch ex Fr.) Karsten (from diminutive of Greek *hydnon*: ancient name for truffle, plus Latin *aurantiacus*: orange-coloured)

In coniferous woods tough leathery or cork-like fungi can be found, with a spiny hymenial surface. They often grow in colonies united at the base, which may be in moss or a carpet of pine needles, or they surround blades of grass, small plants or twigs, enclosing them completely. They can even smother young pine saplings. The spine fungi were placed in the genus *Hydnum*, but their closest relative is *Thelephora terrestris*, which they closely resemble in the way they grow. The genus *Hydnellum* includes about ten species, none of them edible. Those illustrated are some of the commonest and most widespread.

Hydnellum aurantiacum is easily recognized when young by its more or less orange colour. With time the colour becomes more brownish, and it may then be hard to distinguish from related species. The cap is thick and irregular, and its upper surface is lumpy and uneven – first whitish like wool, then orange-yellow. It grows to 10 cm wide. The spines are white at first, then brown with pale tips. The stem is fine and velvety, dark orange-yellow, and 4–8 cm high. The smell is mealy and pleasant. August to October is its growing season.

Hydnellum
aurantiacum *flesh is
divided into zones*

Hydnellum suaveolens ⊗

(Scop. ex Fr.) Karsten (from Latin *suaveolens*: pleasant-smelling)

Easily recognized by its strong, sweet, perfume-like smell, especially marked when the fungus is dried. Before the days of perfumed soaps, this fungus was often put in the linen cupboard instead of lavender. For much of its life the cap is white or yellowish white with a tinge of blue, then later becomes more blue-grey. It has an irregular upper surface which is soon depressed in the middle. It grows to 10 cm wide and the spines are dense and awl-shaped. They are first blue-grey, then whitish, finally purple-brown. The stem, 1–2 cm high, is dark blue, irregular and rather felt-like. Grows mainly on the carpet of fallen needles in pine woods.

Hydnellum suaveolens
*flesh is also divided into
zones with bands of
blue and white*

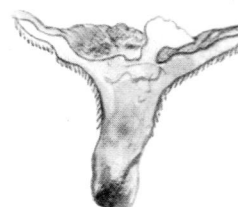

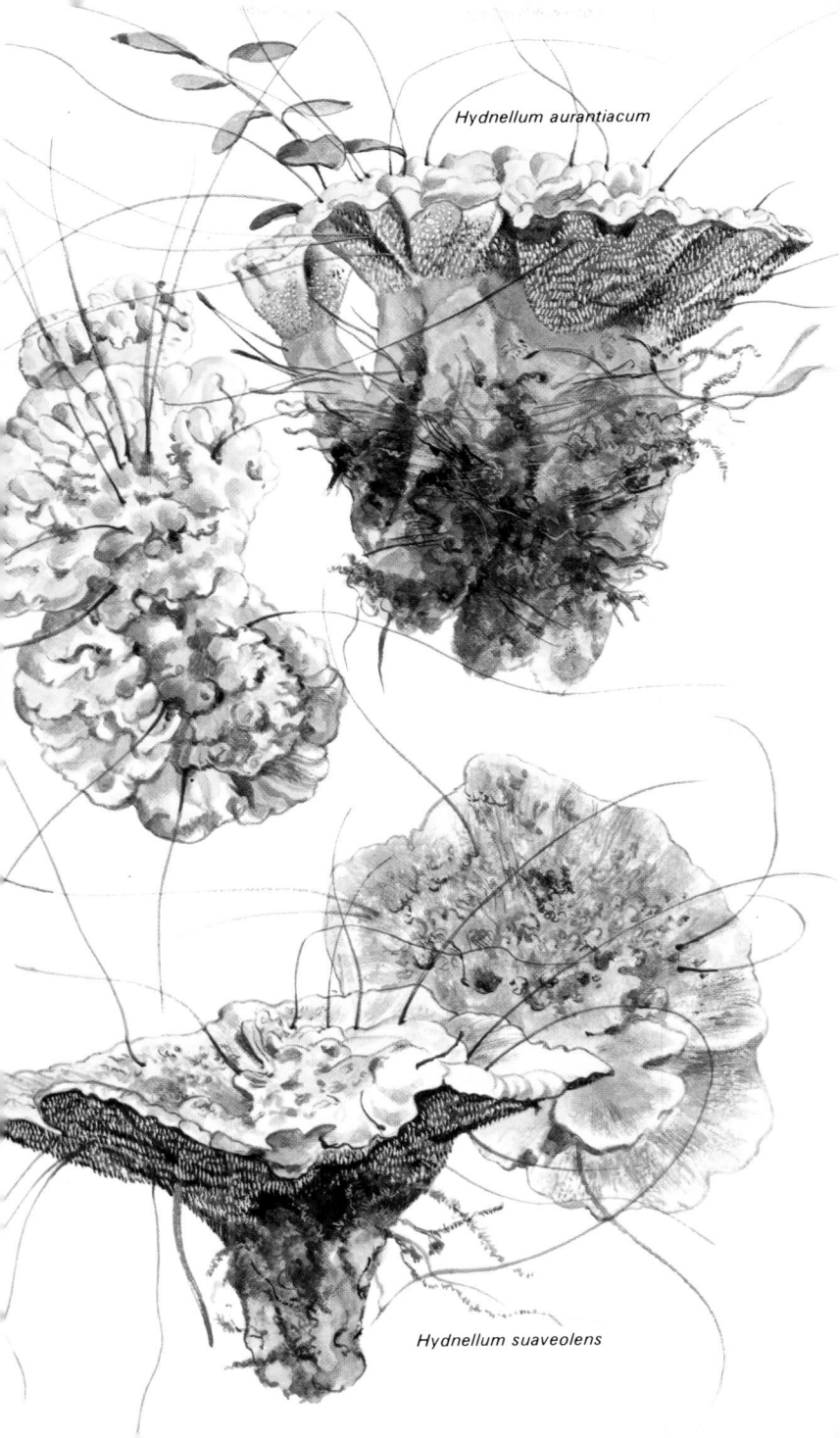

Hydnellum aurantiacum

Hydnellum suaveolens

Polypores

The fungi which go under the general name of polypores are found on living and dead wood, on branches and stumps, and occasionally on the ground. They are usually fairly hard, like wood, cork or leather, with a spore-producing tube layer like the boleti (see p. 100 ff.). However, the tubes of the polypores cannot be freed from the cap as easily as those of the boleti. Elias Fries regarded the polypores as a uniform group, the family of *Polyporaceae* (from Greek *poly*: many, and *poros*: hole, pore) consisting of only a few genera. More recently the polypores have been split into several families and a large number of genera. The characteristics that determine the modern groupings are largely microscopic.

Albatrellus ovinus ⊖

(Schaeff. ex Fr.) Kotl. & Pouzar (from Latin *albatus*: dressed in white and *ovinus*: sheep)

Most polypores are too tough or unpalatable to eat, even though none of them are thought to be poisonous. This ground-growing species is the one truly edible fungus of the group. It grows in groups in mossy conifer woods, especially where the ground is hilly and stony with a sandy base. The caps, white at first and then yellow-grey, often overlap, and their shape becomes bulging and irregular. Their size varies from 10 to 20 cm. The tube layer is white and the flesh is white, but slightly yellowish in older specimens. The white stem is short. The fungus is good to eat and the flesh turns greenish yellow when cooked. Although fairly common throughout most of Europe, this species has not been confirmed in Britain.

Albatrellus confluens (⊖)

(Alb. & Schw. ex Fr.) Kotl. & Pouzar (from Latin *confluens*: flowing together)

In this species several fruitbodies often grow together. The caps overlap like roof-tiles and take on an irregular shape. The pale yellow-red to yellow-brown colour is sometimes reminiscent of slightly creamed wheatbread. The tube layer is white, but turns yellowish red when pressed. The flesh is white, with a bitter taste and a pleasant smell. It can be eaten if well boiled, but it is no delicacy, and there is much to be said for leaving it growing. As with *Albatrellus ovinus* this species is not known in the British Isles.

Albatrellus ovinus

Albatrellus confluens

Laetiporus sulphureus

(Bull. ex Fr.) Bond. & Singer (from Latin *laetus*: happy, lively and *sulphureus*: sulphur)

Sulphur Polypore

This beautiful fungus – sulphur yellow to light brick-red – attacks old oak trees in the summer and early autumn. It badly damages the heartwood of oaks and other deciduous trees, causing red rot, a typical shrinking rot which colours the wood a reddish brown. The trunk is gradually hollowed out as the rotten wood disintegrates into small, brittle, ragged pieces. This split and decayed wood is often seen in the bottom of the trunks of old hollow oaks. The fruitbodies of the Sulphur Polypore do not develop every year. They grow in tile-like clusters with a common base. Polypores of this group can grow very large and weigh up to 10 kg. The tube layer is sulphur yellow with countless fine pores. The flesh is white and juicy but with a sour taste and smell, and cannot be recommended as food. Older specimens become hard and brittle. Fairly common near oak trees, but also grows on other deciduous trees.

Daedalea quercina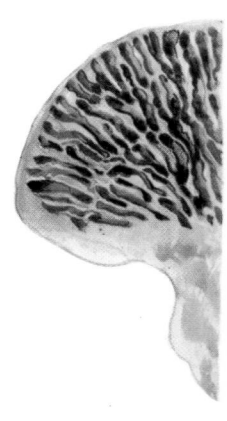

Fr. (from *Daedalus*, who made King Minos' labyrinth, and from Latin *quercinus*: oak)

Another fungus that grows on oak trees. The tube layer resembles the gills of an agaric. It is bracket-shaped, becoming 10–20 cm broad. The yellow-grey fungus has a pleasant smell. It grows mainly on the stumps of deciduous trees, and its distribution is the same as the oak. It is a saprophyte, and therefore grows only on dead wood, but it can damage timber and wooden buildings.

Gloeophyllum sepiarium

(Wulf. ex Fr.) Karsten (from Greek *gloios*: sticky, tough and *phyllon*: leaf, plus Latin *saepiarius*: growing on fences)

An annual fungus, fairly common on conifer stumps and trunks. The fruitbodies are tough like those of *Daedalea* and the tubes have the same gill pattern, but it is brown to rust brown. The margin of the fruitbody is yellowish and hairy. Causes shrinking rot on posts and other similar wooden objects. Very hardy and can stand high temperatures, and even under dry conditions can destroy solid telegraph poles in a very short time.

Laetiporus sulphureus

Gloeophyllum sepiarium

Daedalea quercina

Piptoporus betulinus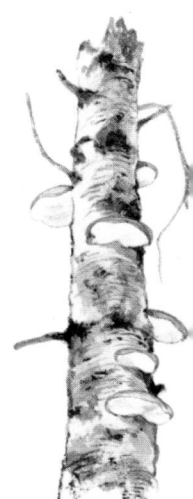

(Bull. ex Fr.) Karsten (from Greek *pipto*: to fall and *poros*: pore, plus Latin *betulinus*: birch)

Birch Polypore

On old and dying birch trees one often sees the cushion-like, kidney-shaped fruitbodies so characteristic of the Birch Polypore. Dead trees may have many of these polypores at different stages of development. The fruitbody appears first as a soft elastic 'bun', which pushes through the bark in late summer or autumn. The full-grown polypore survives to the following spring. The fungus causes a brittle rot which is first pale brown, then yellowish white and the affected wood breaks away easily. In autumn the leaves of living trees with this rot often turn brown sooner than others.

The Birch Polypore can grow to a breadth of 25 cm, but is usually considerably smaller. The upper side is smooth, grey-white to grey-brown, and covered with a thin layer of skin. The edge curls around the tube layer, which is white to yellowy-white in younger specimens, brownish in older ones. Fresh specimens have a rather sour taste and smell. The fungus has been used – and sometimes still is – in several ways. Its Norwegian name (*knivkjuker*) refers to older specimens used as razor strops. The fine mealy fragments left by the rot have been used for polishing in various trades such as clock-making. The Birch Polypore is extremely common and grows mainly in the same areas as the birch tree.

Pycnoporus cinnabarinus

(Jacq. ex Fr.) Karsten (from Latin *pycnoporus*: with compressed pores and *cinnabarinus*: cinnabar red)

Cinnabar Polypore

An attractive orange to bright vermilion polypore. The thin, cork-like fruitbodies, a few centimetres broad, grow on deciduous branches and trunks which have lain on the ground for some time. They are attached by a broad base, and can sometimes be resupinate (see p. 56). The upper side is downy in young specimens, bare in older ones, and is a less brilliant red than the tube layer. An annual fungus which grows on deciduous trees, primarily rowan and birch, it is mostly seen in spring and autumn, and occurs over the greater part of northern Europe, but, apart from a single Scottish record, is unknown in Britain. It causes white rot, but is regarded as one of the less destructive polypores.

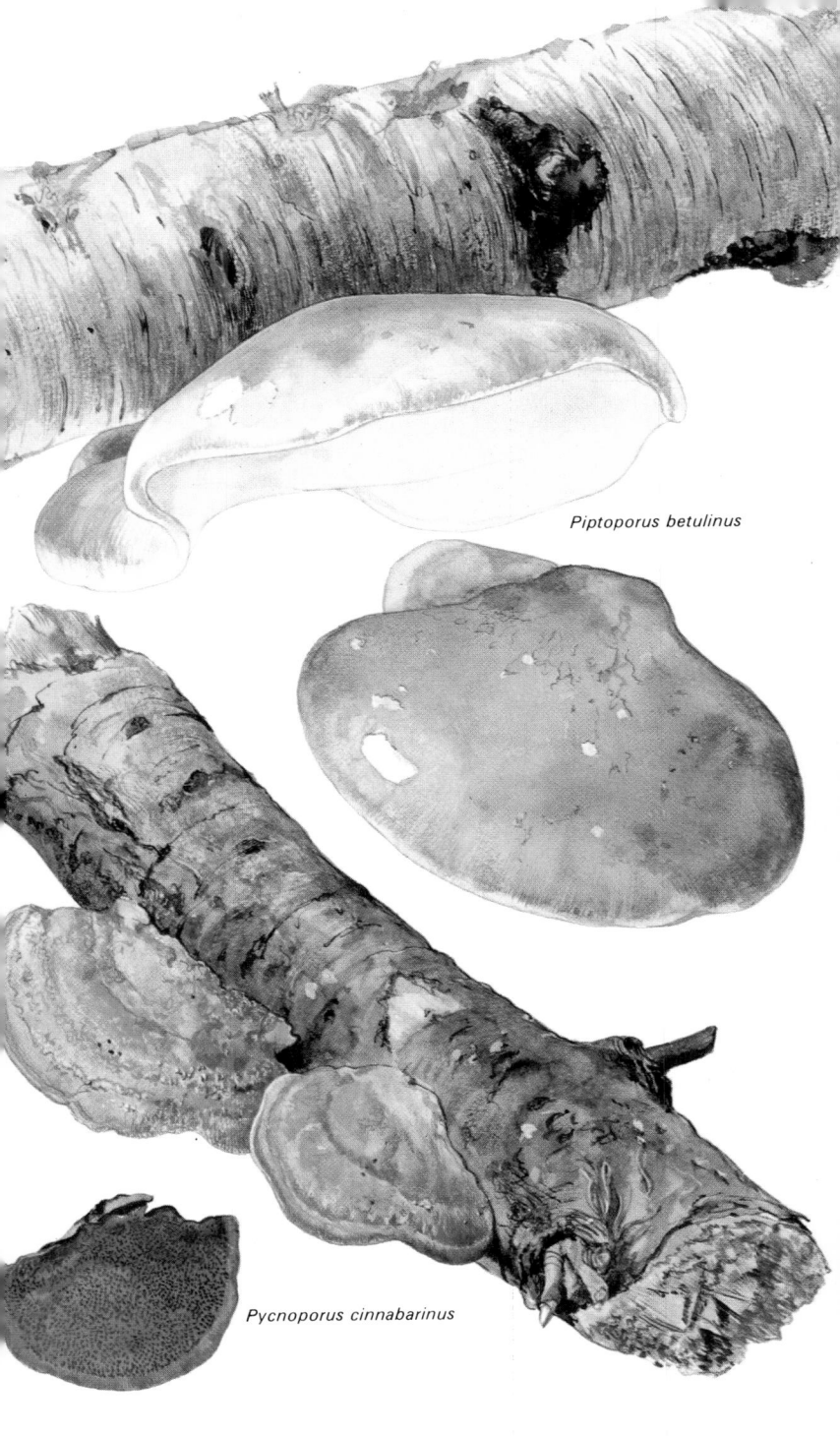

Piptoporus betulinus

Pycnoporus cinnabarinus

Heterobasidion annosum ⊗

(Fr.) Brefeld (from Greek *hetero*: different, *basidion*: basidium, and Latin *annosus*: aged)

Root Fomes

Many polypores attack living trees and cause the highly destructive woodland rot. The most dangerous of the rot-producing fungi is Root Fomes, which directly attacks the roots. Several other fungi may attack roots but the term 'root rot' should be used for attacks by this fungus on the many species of conifer, mainly pine and fir. The fungus can grow on the tree for a long time without killing it, causing great damage in valuable forests. Annual losses from root rot have been estimated at tens of millions of pounds in Sweden alone. The fungus often grows parasitically in the growth layer of a pine, and can kill the tree in a few years. The Root Fomes seems to thrive on chalk, and is common in good woodland soil. It is perennial and wood-like, and often grows pressed against the underside of roots, forming a layer up to fifty cm long. It can also grow in a bracket shape on stumps or fallen trunks. The well-developed fruitbody normally has a relatively thin white margin, and is somewhat furrowed by the zones of each year's growth. At the central point of growth it may be several centimetres thick. The upper side is dirty red to cinnamon brown; the tube layer creamy white to pale yellow. Occasionally the Root Fomes may grow on pine needles and around twigs, as shown in the illustration.

Fomitopsis pinicola ⊗

(Schw. ex Fr.) Karsten (from Latin *pinicola*: growing on pines)

Often grows on fir and pine stumps and although common throughout Europe it is not found in Britain. Like the Root Fomes it causes a brown rot. The wood becomes very brittle and breaks into rectangular pieces. This polypore, however, probably does not do as much damage as other rot fungi. The perennial fruitbody is shaped like a hoof, and is grey to dark brown, with a brilliant yellow-red margin which increases in size. The tube layer is white to sulphur-coloured. The fungus tastes and smells sour. It grows mainly on coniferous wood, but sometimes occurs on the wood of deciduous trees, such as birch.

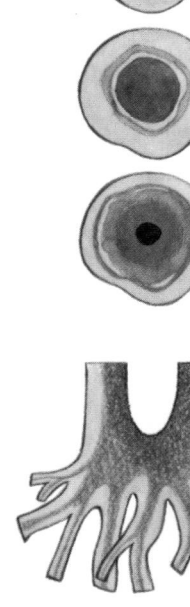

Root Fomes normally infects the tree by damaging its roots, growing towards the heartwood. When the rot is older, it becomes pale brown to red-brown and in its final stages is soft and porous

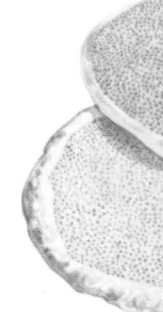

88–89

Heterobasidion annosum

Fomitopsis pinicola

Hirschioporus abietinus ⊗

(Dicks. ex Fr.) Donk (*Hirschioporus* from the name of the scholar R. Hirsch, and from Latin *abietinus*: growing on fir trees)

In relatively undisturbed natural woodland with fallen trunks, one often sees long rows of thin, partly violet-coloured fungi, especially on dead fir trunks and branches. This is *Hirschioporus abietinus*, one of the commonest rot fungi in these areas. It forms a thin leathery fruitbody on vertical surfaces. On the underside of fallen trunks and branches it appears resupinate, in irregular belts that are often a metre long. The upper side of the fruitbody is covered with dense, felt-like hair. It is greyish violet and concentrically furrowed. The tube layer is violet, with a paler shade at the fungus' edge. Older parts of the tube layer become brownish. It causes severe storage rot which begins on the surface and then attacks the sap wood. Affected wood splits easily along the annual growth-rings. Another species is occasionally found on pines. This is *H. fusco-violaceus* ⊗ (Ehrenb. ex Fr.) Donk (from Latin *fuscus*: dark and *violaceus*: violet-coloured) which differs in having the tube layer split into lamellae, forming flaps or spines.

Hirschioporus abietinus *is often covered with algae, and thus becomes greenish*

Coriolus zonatus ⊗

(Nees ex Fr.) Quél. (from Latin *corium*: hide, skin and *zonatus*: zoned)

This polypore belongs to a group of leathery species whose fruitbodies have clear radiating zones. It usually grows on dead trunks, branches and stumps of deciduous trees, especially birch. It causes a white rot and can greatly damage stored timber. The annual fruitbodies are thin and semi-circular, with a breadth and length of 5–10 cm. Several fruitbodies often grow together in a tight group. The upper side of the polypore is downy with concentric grey to brown bands and is usually slightly yellowish at the base. The tube layer is first white, then grey to greyish brown. *Coriolus versicolor* ⊗ (L. ex Fr.) Quél. (from Latin *versicolor*: motley), the Many-zoned Polypore, a closely-related species with an almost silken surface, is probably the most common polypore in the British Isles. They form tight clusters of zoned, brown, velvety fruitbodies, with a white margin and a white tube layer.

irschioporus abietinus

Coriolus zonatus

Ganoderma lucidum

([W. Curtis] Fr.) Karsten (from Greek *ganos*: shining and *derma*: skin, plus Latin *lucidus*: shining)

Species of *Ganoderma* differ from other polypores, partly by the upper surface of the fruitbody having a hard crust, and partly in the appearance of the spores. They belong to the mainly tropical genus *Ganoderma*.

Ganoderma lucidum is a beautiful fungus, easily recognized by its shining red-brown fruitbody, which almost seems to have been varnished. It is hard and cork-like, and can grow up to 20–30 cm in width. A well-defined stem runs obliquely downwards from the back edge of the polypore. The tube layer is white to pale cinnamon-brown. The fruitbody is usually annual, sometimes two or three years old, and grows saprophytically on a wide range of deciduous trees, more rarely on conifers. The fruitbodies appear in late summer and early autumn. The fungus causes a white rot with flecks of black.

Ganoderma applanatum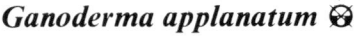

(Pers. ex Wallr.) Pat. (from Latin *applanatus*: flattened)

The flat fruitbodies of this fungus are often found on old stumps. They are rounded or kidney-shaped, and look as if they have been dusted with cinnamon, owing to a covering of spores. The spores, which fall straight down from the tubes, are caught by warm air currents rising from the ground, especially on still nights. The air is stationary above the fungus, and the spores thus fall onto its upper side. The great enlargement of the spore-producing surface makes the spore production very large. This may continue for four to five months, and the total number of spores spread can therefore be as many as 5–10,000 million for each fungus!

The perennial fruitbody is thin, corky, bracket-shaped, and up to 30 cm broad. Several polypores may grow as one formation. The brownish grey to dark brown fungus causes a soft white rot but is not regarded as one of the most highly destructive. It grows mainly on the dead stumps of deciduous trees, and less often on conifers. *Ganoderma adspersum* (Schulz) Donk (from Latin *adspersus*: scattered) is perhaps much more common in Britain. It forms thicker fruitbodies with a darker brown flesh.

Ganoderma lucidum

Ganoderma applanatum

Ischnoderma resinosum ✪

(Fr.) Karsten (from Greek *ischnos*: dry and *derma*: skin, plus Latin *resina*: resin)

Many of the generic names for various groups of wood polypores are the work of the Finnish botanist P. A. Karsten, one of the great nineteenth-century researchers. The special feature of *Ischnoderma resinosum* is that the cap is ultimately covered with a hard resinous film. At first it is fleshy, then corky or leathery and mussel-shaped, with a coarse floccose surface. It is dark rust to coffee-coloured, with barely perceptible belts of shiny black, or sometimes mauve. It grows to 15 cm in width. The margin of the cap is thin, turned downwards or inwards, and in older fungi it is wrinkled. The tube-layer is pale, and the pores are small and whitish, turning brownish when touched. This annual fungus has a pleasant smell, and grows in a random pattern on conifer stumps, but rarely on beech and oak, throughout northern Europe. It causes white rot.

Coltricia perennis ✪

(L. ex Fr.) Murr. (from *coltrichione*: the Italian name for certain fungi, and Latin *perennis*: durable)

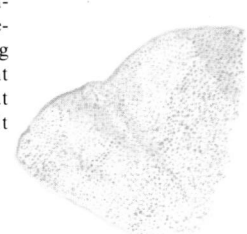

Tube layer of Coltricia perennis

Linnaeus noted and described this easily recognizable polypore, with the thin, tough, skin-like brownish cap – very distinctly zoned – and the centrally-placed stem. It is one of the few polypores which grow in the ground, in sandy pine woods throughout northern Europe. It is now placed in a small genus, *Coltricia*. In Europe this includes one other species, the very rare *Coltricia cinnamomea* ✪ (Jacq. ex Pers.) Murr. (from Greek *kinnamomon*: cinnamon). This is also brown, but is less distinctly zoned.

The cap is at first funnel-like, with fine velvety hairs. Later it flattens out and becomes smoother, growing to a breadth of 3–10 cm. The tube layer is first whitish, then yellow-brown, and the pipe openings are small and angular. The stem is firm, velvety in texture, 2–4 cm high, and merges directly into the cap. This fungus has no appreciable taste or smell. It grows in late summer and autumn, but the dry, tough fruitbodies may survive for several years – hence the Latin name.

hnoderma resinosum

Coltricia perennis

Phellinus igniarius ✪

(L. ex Fr.) Quél. (from Greek *phellos*: cork, and Latin *ignis*: fire)

The Latin name is a reminder that polypores were once used as fuel. This hard, dark-coloured, heavy hoof-shaped fungus burned long and slowly. It is perennial and can live to a great age. The fruitbodies may grow to a very large size – 2–3 cm or more in diameter. It has a wood-like consistency, is coarsely furrowed, and often full of cracks. The upper surface is grey-black to black and grey-brown to rust on the tube layer of the underside. It is parasitic, penetrating through 'wounds' in the tree. The spores infect the dry and damaged branches of deciduous trees, mainly birch and aspen, and then invade the heartwood of the trunk. *Phellinus igniarius* is cushion-shaped or spread out on the substratum, and the fruitbody has a rounded edge. It is brown to brown-grey on the upper surface, yellow-brown to dark brown on the tube layer. A number of closely related species specific to certain kinds of wood have now been recognized. *Phellinus nigricans* ✪ (Fr.) Karsten (from Latin *nigricans*: blackening) is specific to birch. *Phellinus pomaceus* ✪ (Pers.) Maire (from Latin *pomum*: tree-fruit) was previously regarded as a sub-species of *P. igniarius*. It very often grows on ageing plum trees, usually causing their death.

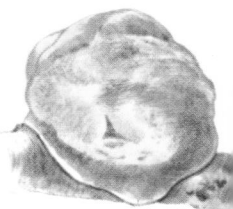

Phellinus pomaceus

Fomes fomentarius ✪

(L. ex Fr.) Kickx. (from Latin *fomes*: tinder)

Tinder Fungus

Probably the best-known polypore. Together with *Phellinus igniarius* it was an important source of tinder and is still used in some countries. The cork-like fabric was put into an alkaline solution to soften it, and then dried and pounded into thin flakes. It was also used for making garments, boxes and ornaments. It is perennial, hoof-shaped and strongly coloured, often very large with the fruitbodies often growing together to form groups. There is a soft corky flesh, and the fruitbodies are light. The tube layer is cinnamon brown. It causes a white rot, mainly on beech and birch. It is fairly common on birch in Scotland, but only occasionally found in southern England, where it grows on beech.

Vertical section of Fomes fomentarius

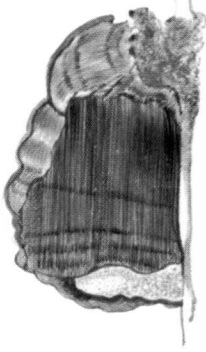

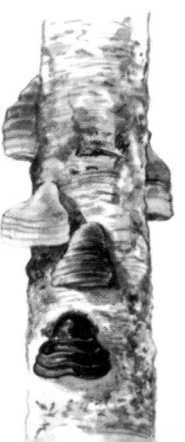

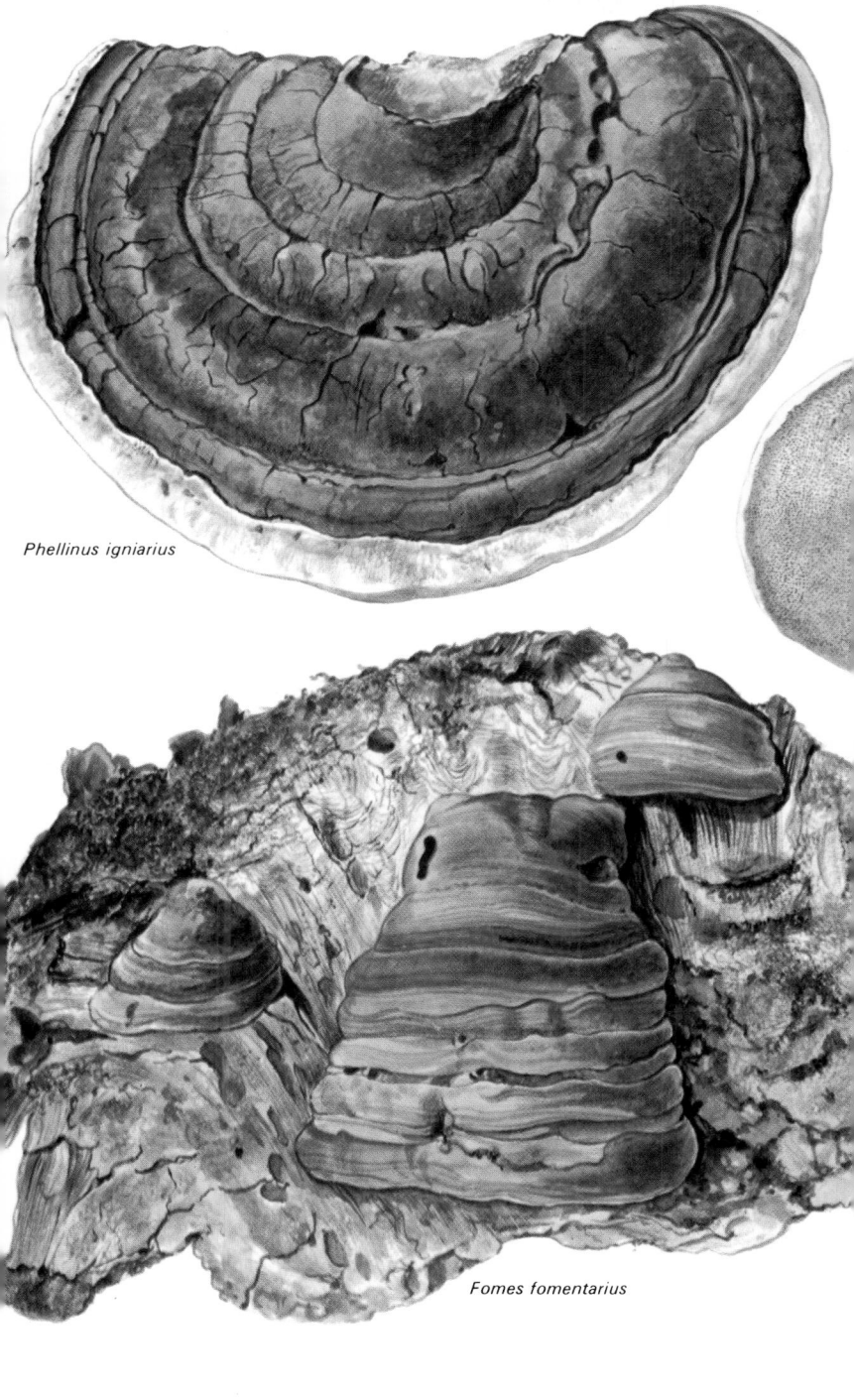

Phellinus igniarius

Fomes fomentarius

Polyporus squamosus ⊖

(Huds. ex Fr.) Fr. (from Latin *squamosus*: rich in scales)

Scaly Polypore, Dryad's Saddle

An annual fungus with a distinct cap and stem, it really belongs to the genus *Polyporus*. Because of the relationship to *Pleurotus* species, some scholars classify the genus *Polyporus* and the family *Polyporaceae* under the order Agaricales which include the agarics.

The Scaly Polypore has a kidney- or fan-shaped cap, which may be 50 cm or more in width. The cap is whitish yellow to ochre, with dark concentric scales. The stem is white at the top with a black base. The tube layer is white to pale gold. Young specimens are edible. The fungus grows on stumps or on living trees, mainly on maple, elm, ash and cultivated fruit trees, and causes a serious white rot. In several countries, especially Bulgaria, the mycelium of this species has been cultivated commercially for, among other products, molasses.

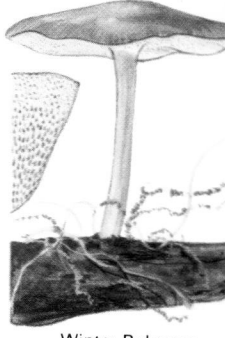

Polyporus brumalis ⊗

(Pers. ex Fr.) Fr. (from Latin *brumalis*: winter)

Winter Polypore

Winter Polypore

A slender and graceful polypore found in late autumn and winter on the stumps of deciduous trees, and on trunks and branches, often hidden in the ground. The fruitbody, which resembles an agaric, has a rounded cap and a centrally-placed stem. The cap is grey-brown and grows to 8 cm in width. The tube layer is white, and the pores are up to 1 mm. The stem is brownish. It occurs fairly generally in large areas of northern Europe.

Fistulina hepatica ⊙

(Huds.) Fr. (from Latin *fistula*: pipe and *hepaticus*: liver)

Beefsteak Fungus

Beefsteak Fungus flesh resembles muscle tissue or tough fibrous meat

This representative of the genus is one of the most remarkable of the polypores. The fruitbody, looking like a liver or a tongue, is easily recognized and has a characteristic habitat: on old, knotted oak trees. When young it is damp and slimy but in older specimens the tubes are easily separated (see illustration, lower right next page). The fungus secretes a wine-red juice. The taste is somewhat sour and the fruitbody is edible when young, but needs to be soaked carefully because it contains tannic acid. The Beefsteak Fungus occurs where there are oak and beech trees.

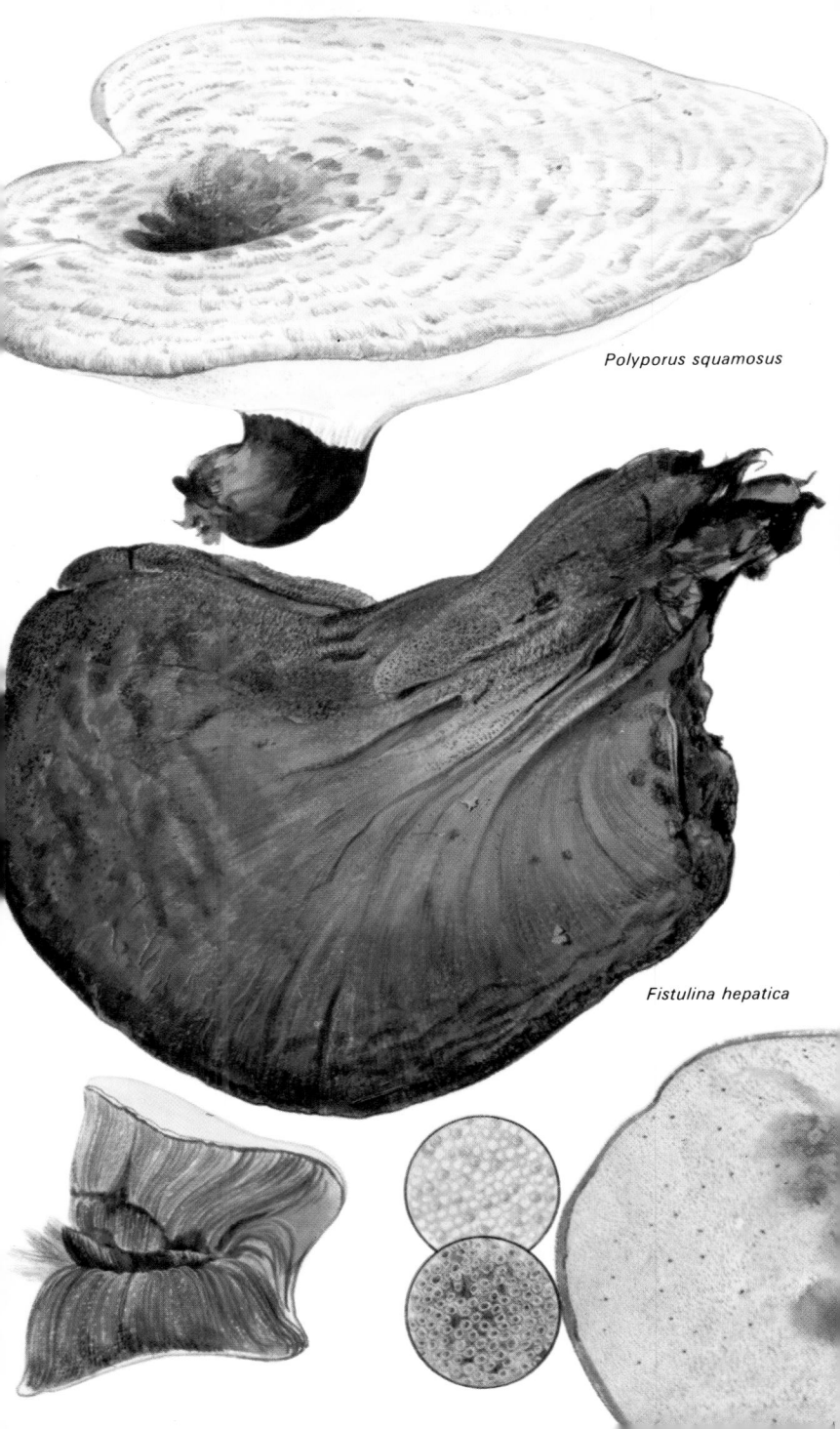

Polyporus squamosus

Fistulina hepatica

Boletus edulis ⊖

Fr. (from Greek *bolos*: lump, and Latin *edulus*: edible)

Penny Bun Bolete – Cep

The Penny Bun Bolete is a general name for several boleti,
all of which form mycorrhizal associations like most other
boleti (see p. 18). Certain species of the *B. edulis* complex can
grow with a variety of deciduous and coniferous trees,
while others grow with only one kind. Boletes multiply
rapidly in a short time when it is warm and rainy – a
'boletus year'. Of the species illustrated, *Boletus aestivalis* is
an early summer fungus, confined to deciduous woods, and
Boletus pinicola is found in late summer in coniferous woods,
mainly in Scotland. Little is known about the distribution of
B. aestivalis but it is particularly widespread in England under
beech and oak. *Boletus edulis* can grow both in deciduous and
in coniferous woods. Young specimens of *Tylopilus felleus*
(see p. 102) can be confused with the Penny Bun Bolete.

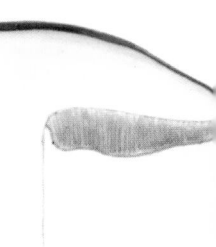

The flesh of Boletus
edulis *is white, except
just below the skin of
the cap, where it is
brown to violet*

The cap of *B. edulis* is date brown to chestnut brown, or
paler. It is convex at first, then almost flat. The rim is slightly
lighter in colour, and the fungus grows up to 20 cm wide. The
cap skin is often rather uneven, and when damp it is shiny
and slightly sticky. The tubes are tightly packed, at first
grey-white, then yellow to olive green. The light grey to light
brown stem has a fine network of pale veins. At first it is
often swollen at the base; later it is more even in thickness,
and grows to a height of 15 cm. With its pleasant smell and
nutty flavour, it is one of the most sought-after edible mushrooms, particularly prized in France. *Boletus edulis* grows
during August and September.

The very similar *Boletus pinicola* ⊖ (Vitt.) Venturi (from
Latin *pinicola*: growing on pine trees) has an almost copper-
coloured cap. The tubes soon become yellow-green to olive,
with rust-coloured openings. The stem is darker than in
B. edulis.

Boletus aestivalis ⊖ Fr. (from Latin *aestivalis*: summer) has
a mainly grey-yellow to light brown cap and suggests chamois
leather. In dry weather the skin of the hat often cracks into a
mosaic of squares. This fungus has the alternative Latin name
Boletus reticulatus Boudier (from Latin *reticulum*: little net).
Like others of the group this species makes excellent eating.
A great favourite with squirrels, it sometimes grows profusely
in oak and beech forests, and also in birch groves.

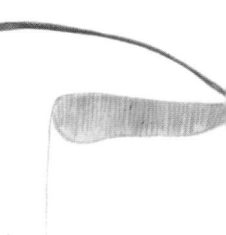

Boletus aestivalis *has
white flesh which
extends just below the
skin*

etus edulis

Boletus pinicola

Boletus aestivalis

Tylopilus felleus ✪

(Bull. ex Fr.) Karsten (from Greek *tylos*: bump and *pilos*: hat, plus Latin *felleus*: gall)

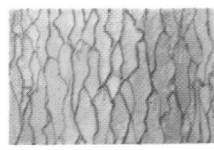

This is one of the boletes which may grow with both deciduous and coniferous trees, forming mycorrhizal associations with them. It is widely distributed in Britain in oak and beech woods, but is more commonly found in England. It may grow in the company of the Penny Bun Bolete with which it can be confused when young. Sometimes it can be found on decaying stumps. It is therefore doubtful whether it always forms mycorrhizal associations. In chalky districts it grows mostly where the chalk is waterlogged or where the fallen leaves are thick.

Young Penny Bun Boletes have a faint, pale, fine-meshed network of veins on the stem. The veins of a young Tylopilus felleus *are clearly marked, coarse-meshed and brown*

The majority of boleti were formerly included in a single genus, *Boletus*. They have now been divided into several genera. *Tylopilus felleus* is characterized by the pink colour of the spores. It is the only European species within this genus, which has some twenty species altogether, some of them edible and mildly flavoured. *Tylopilus felleus* is normally somewhat smaller than the Penny Bun Bolete and grey-yellow to pale brown on both cap and stem. The cap is first convex, then almost flat, and up to 15 cm wide. The cap skin is dull or only slightly shiny, with a velvety surface in dry weather. The fleshy stem is paler on the top than at the base, which is somewhat dilated in young specimens. It grows up to 10 cm tall and 4 cm thick. The flesh is white and reddens slightly. Although the scent is pleasant, the taste is bitter and it cannot be called edible – a good mixed mushroom dish can be ruined by a single one. To be quite sure of avoiding this, there is no harm in tasting a little bit of the flesh: it is absolutely harmless. It grows fairly generally during August and September.

Older Tylopilus felleus *tubes are pink with the ripe spores, while those of older Penny Bun Boletes (see below) are yellowish*

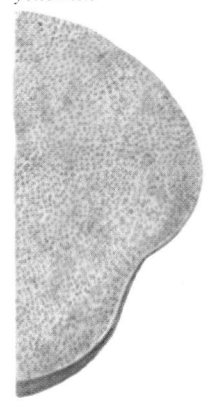

Tylopilus felleus

Boletus luridus ⊗

Fr. (from Latin *luridus*: sallow)

In beech, oak and lime woods, especially on chalky soil, there are various kinds of attractively coloured boletes. They have thick stems, veined in varying degrees, and bright red pores. Many also have a reddish stem, and the flesh often turns blue when broken. They appear in great profusion after warm summer rains. Some need a great deal of warmth, and appear only in certain congenial places in certain years. Others are found here and there nearly every year. One of the latter is *Boletus luridus*, found on chalky ground in deciduous woodland. Several of these pretty reddish boletes are thought to be poisonous.

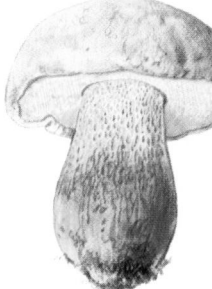

Boletus calopus

The three main characteristics of *Boletus luridus* are flesh which quickly changes colour to greyish violet or blue-green when broken; the dark, distinct, coarse-meshed network of veins on the stem; and the rusty-red pores. The cap is fleshy, first convex, later expanding, and grows up to 12 cm wide. It is grey-yellow to olive brown, with a pink tinge at the edge. In young specimens the stem is distended at the base. This fungus has a sourish smell and taste and has been described as edible, but because of the risk of confusing it with poisonous species it should be avoided. The fruitbody grows from June to September in parks and avenues of limes.

Boletus satanas

One of the prettiest boletes is *Boletus calopus* ⊗ Fr. (from Latin *calopus*: with a pretty foot). Its cap is sallow to olive grey, up to 12 cm wide, and the tubes are first sulphur yellow, then greenish. The chunky stem is yellow above and carmine below, with a distinct net of pale veins, and can grow to 10 cm tall. The flesh of both cap and stem is white, and turns blue when broken. The scent is ink-like and the taste bitter. It grows on chalky ground from July to September, in both deciduous and coniferous woods, but especially under beech and oak.

Boletus satanas ⊗ Lenz (from Hebrew *satan*: devil). The Devil's Bolete is perhaps the bolete that has attracted most attention. This is because of its size (it can grow to a width of 20 cm), its pale whitish cap, bright red stem and red pores. The Devil's Bolete requires chalky ground and a lot of warmth, and is uncommon. In Britain it occurs mainly in the south of England.

Boletus luridus

Boletus subtomentosus ☉

Fr. (from Latin *subtomentosus*: fine-haired)

Yellow-cracked Bolete

This is one of a group of boletes whose caps have a dry felt-like or velvet-like skin which is very hard to remove. It is sometimes placed in a genus of its own, *Xerocomus* (from Greek *xeros*: dry and *kome*: hair). It does not seem to be specific to any particular type of tree, and some investigations suggest that it can grow either by forming mycorrhizal associations or by breaking down leaves and pine needles. This fungus varies a good deal and can be subdivided into several forms. It is common in frondose and mixed woodland. On moorland there is a large form with a short stem (see top illustration opposite).

The cap of the Yellow-cracked Bolete is convex, olive yellow to olive brown, always dull and velvety, and up to 12 cm in width. The tubes are first pale yellow to gold, then olive. The stem is often of uniform thickness, yellowish to yellow-brown, and sometimes has slight longitudinal ribbing. It grows up to 10 cm tall. The flesh of both cap and stem is white to yellowish (see top illustration this page). The scent is slight, the taste mild and it is a good ingredient for a mixed mushroom dish. It is usually found from July to October.

Another member of the same group is the Red-cracked Bolete *Boletus chrysenteron* ☉ Fr. (from Greek *krysos*: gold and *enteron*: entrails). Its habitat is deciduous woods, and it is common in beech woods, in parks and among clumps of oak trees. It is easily distinguished from the Yellow-cracked Bolete by the pale red flesh under the cap skin, which often shows through the chequered cracks. Grows from July to September.

Cross-sections:
Yellow-cracked
Bolete (top), Boletus
piperatus *(centre) and*
the Red-cracked Bolete

Boletus piperatus ✪

Fr. (from Latin *piperatus*: peppered)

Easily recognized by its strong peppery taste, cinnamon-coloured cap and stem, its rust-coloured tubes and the lemon to chrome flesh at the stem base. The cap is shiny when dry and is 2–8 cm in width. The slender stem grows up to 8 cm. The scent is slight. Widely distributed, this small fungus grows in both deciduous and coniferous woods.

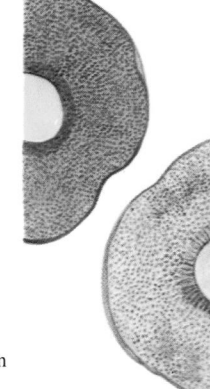

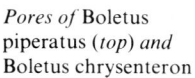

Pores of Boletus
piperatus *(top) and*
Boletus chrysenteron

106–107

Boletus subtomentosus

Boletus piperatus

Boletus chrysenteron

Boletus badius ☉

Fr. (from Latin *badius*: chestnut brown)

Bay-coloured Bolete

Like *Boletus subtomentosus* this species seems able to live
either as a mychorrhizal fungus or by breaking down dead
matter such as pine needles, leaves and twigs. It is often
found among fallen pine needles, but also on rotten stumps
long after the whole wood has been felled. It grows both in
poor and in chalky soil.

This fungus is fairly easy to recognize by its chestnut brown
cap (sticky in wet weather, shiny in dry) and the dirty yellow
pores which turn blue when touched. In other ways it is
similar to *Boletus subtomentosus* to which it is closely related.
The cap is at first convex, then almost flat, and grows up to
12 cm broad. The stem is 8–12 cm tall, somewhat lighter
brown than the cap, often of uniform thickness, and sometimes has slight ridges. With its slight scent and mild taste,
it is almost as good to eat as the Penny Bun Bolete. Found
during August and September.

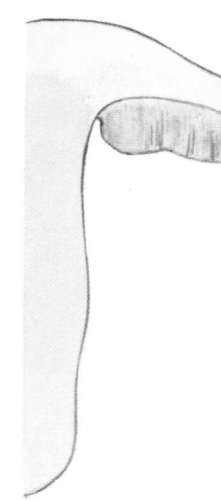

*The yellow-pink flesh
of* Boletus badius *turns
blue-green when
broken*

Suillus granulatus ☉

(L. ex Fr.) O. Kuntze (from Latin *suillus*: pig and *granulatus*:
with small grains)

Found in sandy pine woods from early summer until well into
the autumn. It is yellowish to yellow-brown, with a sticky or
slimy cap and dense light yellow pores, which in damp
weather exude milky drops.

This species, together with other boletes with slimy caps,
are now placed in a special genus *Suillus* which includes about
fifteen species in Europe. They are among the most important
mychorrizal fungi of the pine woods. The cap starts off
convex, then spreads into a cushion-like shape, and grows to
10 cm in width. The skin of the cap comes away easily. The
stem is whitish to pale yellow, 4–8 cm tall, and the upper part
has floccose granules which gradually darken. The scent is
slight and the taste mild, and it is good to eat.

Suillus granulatus
flesh is very pale yellow

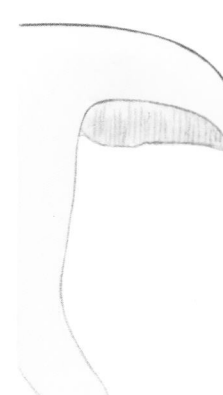

Boletus badius

Suillus granulatus

Suillus luteus ☉

(L. ex Fr.) S. F. Gray (from Latin *luteus*: yellow)

As with the related *Suillus granulatus*, this species is always associated with coniferous woodland, especially Scot's Pine. Found over large areas of the world in sandy soil, even in places where the pine is not indigenous, it is uncommon where there is no snow. In Britain it is widely distributed and common. It is certainly one of the pine's most important mychorrhizal fungi, but probably also grows in association with firs. Typical of *Suillus* species, the cap has a sticky or slimy skin easily pulled off. It also has a ring on its stem, a feature of several other *Suillus* species. Highly regarded as food, but very prone to attack by insect larvae in warm weather. In the late autumn, however, when the insect breeding season is over, there are often rich harvests of 'maggot-free' specimens.

The generic name *Suillus* now applied to a particular group of boletes goes back to Roman times. The boletes were called *fungi suilli* (pig fungi). In the country nowadays they are known as 'cow mushrooms'.

The cap of the *Suillus luteus* is often chocolate brown to olive brown, convex, and up to 12 cm wide. The small pores are lemon yellow at first and then brownish. The stem is yellowy white with wine-coloured granules, and grows to 5–10 cm tall. The ring, which in young specimens joins the margin of the cap to the stem, is at first white, but is gradually darkened by the spores and falls down along the stem. In old specimens it often shows only as a dark band around the stem. The flesh is yellowish white. The scent is slight, the taste mild and it often grows along paths and woodland roads from July to October.

Suillus luteus

Suillus grevillei ⊖

(Klotzsch) Singer (after the Scottish botanist R. K. Greville)

Larch Bolete

Several mycorrhizal boletes are restricted to particular species of tree. One of these is the Larch Bolete, which until well into the nineteenth century was confused with *Suillus luteus*. It seems to follow the larch's area of distribution. In Sweden the larch was first planted on a large scale in the 1840s. The Larch Bolete was then rare, but around 1860 Elias Fries – the first man to see a clear connexion between the larch and the Larch Bolete – noticed that this fungus was becoming more common. It grows throughout the range of the host tree, whether the soil is rich in chalk or not. There is no certain proof that the fungus can form fruitbodies without the larch.

The Larch Bolete closely resembles *Suillus luteus*, but it has a lighter-coloured cap and there are no granules on the stem. The pores are sulphur yellow and wider than those of *Suillus luteus*, and red-brown patches appear on them when touched. The cap is gold-yellow to yellow-brown, up to 10 cm wide. The stem is darker than the cap, 6–12 cm tall, and has a white to yellowish ring, soon torn and coloured olive by the spores. This fungus has a slight scent, a mild taste and is good to eat. It grows in August and September.

Gyrodon lividus ⊗

(Bull. ex Fr.) Sacc. (from Greek *gyros*: whorl, and Latin *lividus*: lead grey)

This species is wholly restricted to the alder. It grows in alder carrs, often hidden in the grass, and it may be widely distributed although rare. Probably it is simply overlooked, although a chalky soil may be a prerequisite. It is the only species of the genus in Europe, and is clearly distinguished from other boleti by the short tubes with irregular openings running some distance down the stem.

The cap is a muddy straw-coloured or pale ochre to cinnamon up to 10 cm broad and varies from convex to flat. It is sticky when damp. The pores are sulphur yellow to olive, and turn blue when bruised. The stem is the same colour as the cap at first, but soon darkens. Like the other parts of the fungus it acquires brown streaks where touched. The stem grows up to 8 cm high. The scent and the flavour are both slight. It grows during August and September, and is not edible.

Gyrodon lividus has pale yellow flesh, which turns blue in patches, and is rust-coloured towards the base

Suillus grevillei

Gyrodon lividus

Suillus variegatus ⊖

(Sow. ex Fr.) O. Kuntze (from Latin *varius*: motley)

Grows in damp coniferous woods and pine plantations and is widely distributed and common in Scotland and England. It belongs to the same group of boleti as the sticky varieties, with an easy-to-remove skin on the cap, but it is sticky only in damp weather. In dry weather the skin is felt-like and cracked into tiny scales. Other features of *Suillus variegatus* are the small, sand-coloured pores, the ringless stem and the yellow-brown colour.

The cap is at first almost hemispherical, then convex, with rust-coloured scales which gradually darken and finally disappear. It is 5–12 cm broad. The pores are olive to brownish and the stout firm stem is pale yellow to ochre, sometimes with a reddish tinge. It grows up to 10 cm in height. The pale yellow to faint orange flesh sometimes develops blue patches when cut. The faint scent and mild taste make it a good ingredient for a mixed mushroom dish. Often grows profusely from August to October and is not so readily attacked by insects as other boletes.

Suillus bovinus ⊖

(L. ex Fr.) O. Kuntze (from Latin *bovinus*: cow)

Widely distributed and common, this species is characteristic of coniferous woods and plantations. Found both in chalky and in poor soil, it occurs in the Alps and may be found up to 800 m above sea level, but it is reported to be rare at such heights.

Suillus bovinus is easily recognized by the slimy grey-yellow to grey cap, the pale yellow-green to olive grey pores and the ringless stem. The cap is convex, up to 15 cm wide. The stem, often the same colour as the cap, tapers somewhat towards the base. It is 4–6 cm tall. The faint yellow-pink flesh darkens towards the base of the stem. The scent is faint and the taste mild, and it is an edible species.

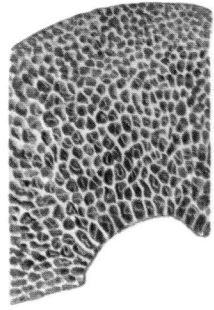

The large angular pores of Suillus bovinus *are subdivided into smaller pores with sunken dividing walls*

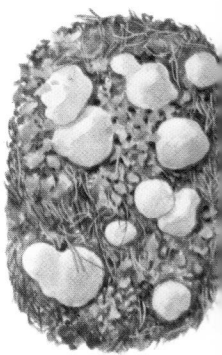

Suillus variegatus

Suillus bovinus

Leccinum versipelle ☉

(Fries & Hök) Snell (from Italian *leccino*: a type of rough-stemmed boletus, and Latin *versipellis*: which changes its skin)

Various names have been given to orange- or brick-coloured boletes with tufts of scales on their stems and flesh that darkens when broken. Recent study of these fungi and their habitat has made it possible to distinguish a number of species. All of them seem to be mycorrhizal and most are linked to particular types of deciduous trees.

The genus *Leccinum* – the 'rough-stemmed boleti' – are clearly distinguished from other boletes by the tufts of scales on the stem. In Britain there are thirteen species. *Leccinum versipelle* differs from the other species in that young specimens have crowded black scales on the stem. The cap is yellow-brown to orange, first hemispherical, then expanding, and up to 20 cm wide. The robust stem can grow as high as 20 cm. With time the scales separate and become more distinct. The pores are initially mouse grey fading to ochreous. The flesh becomes wine red when broken, then grey-lilac to grey. The scent is slight and the taste mild. This species makes excellent eating, and mixes well with other fungi such as the Chanterelle. It grows under birch trees between August and October, very profusely in some years.

Leccinum aurantiacum ☉

S. F. Gray (from Latin *aurantiacus*: fire-coloured, orange)

The name *Leccinum aurantiacum* is often used for all orange-coloured 'rough-stemmed boleti'. In fact it differs greatly, especially when young, from the other species. It grows only near the aspen in good soil (preferably containing chalk) and is thought to be rare in southern areas. The tubes, like the stem scales, are almost white at first. The convex cap is yellow-brown to red-brown, and up to 20 cm in width. The stem is up to 15 cm tall. With time the scales become red-brown to dark brown. The scent is faint and the taste mild. It often grows along grassy verges and among decaying leaves from July to October.

A species that has recently been separated from *Leccinum aurantiacum* is *Leccinum quercinum* ☉ (Pilát) Green & Watling (from Latin *quercinus*: oak). It has a darker red-brown cap, and even when young has a red speckled stem with red-brown scales. It usually grows in beech and oak woods.

In the mature pine woods of Scotland there is a copper red species *Leccinum vulpinum* ☉ Watling (from Latin *vulpinus*: fox).

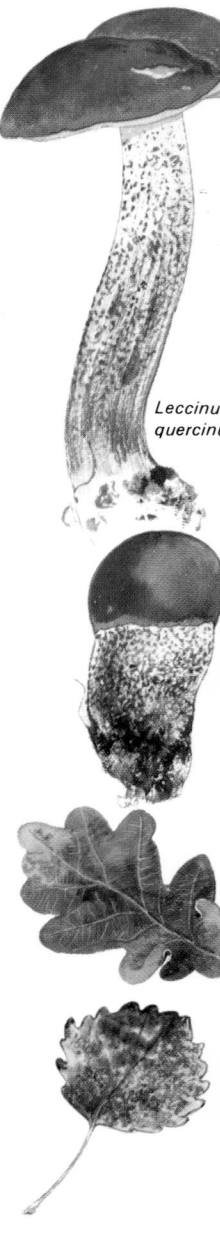

Leccinu[m]
quercin[um]

Leccinum versipelle

Leccinum aurantiacum

Leccinum scabrum ⊖

(Fr.) S. F. Gray (from Latin *scaber*: rough)

Brown Birch Bolete

People used to distinguish between two types of 'rough-stemmed boleti', *Leccinum aurantiacum* and *L. scabrum*. The first was for stoutly-built orange-coloured ones, the second was for the slimmer brown ones. Like the orange species (see previous entry) the brown ones can be subdivided into several species, although in this case they are a good deal harder to separate. The majority grow under various types of birch. The commonest species have a brown to grey-brown cap, and flesh which does not change colour when broken. This is the one illustrated, called *Leccinum scabrum*. Another species, *Leccinum roseofracta* ⊖ Watl. (from Latin *roseus*: pink and *fractus*: broken), perhaps with a northerly distribution, has an almost black cap with lighter patches, and flesh which becomes slightly pink when exposed. It is more robust and grows in damp conditions, especially among peat moss. *Leccinum holopus* also likes peat moss. Both species shown are found throughout Britain. There are also some brown 'rough-stemmed boleti' whose flesh darkens when cut. One of these, *Leccinum carpini* ⊖ (R. Schulz) Moser ex Reid (from *carpinus*: hornbeam) is specific to hazel and hornbeam and comes at midsummer. It is more common in southeast England. Some other species with flesh which changes colour are related to the orange-coloured 'rough-stemmed boleti'.

The cap of the Brown Birch Bolete is first hemispherical and hard, then convex and soft, 5–15 cm in width. The small pores are at first whitish, then grey to brown-grey, and shorter close to the stem. The stem is white to greyish, with small scales which are cigar brown to black. It can grow as tall as 20 cm. The scent is faint and the taste mild. Young specimens are good to eat. This fungus grows from June to October in various kinds of soil and is often found in grass.

Leccinum holopus ⊖ (Rostkovius) Watling (from Latin *holo*: entire and *pes*: foot) is distinguished from the Brown Birch Bolete partly by the lighter colour of its cap, which is sometimes almost white, often tinged with green. The flesh is white, but acquires blue-green patches at the base of the stem when the fungus is cut. It grows in marshes with peat moss, and may have a long stem sunk deep into the moss.

Brown Birch Bolete is often attacked by insects which lay their eggs in the tube layer, and then the larvae spread into the fungus. The hatched insects lie in hollows in the cap's flesh, close to the tube layer

Leccinum scabrum

Leccinum holopus

Paxillus involutus 🕸

Batsch ex Fr. (from Latin *paxillus*: peg, plug and *involutus*: turned inwards)

Brown Roll-rim

The illustrations opposite show two fungi with gills on the underside of the cap. It was thought that all such fungi belonged together but new research has revealed that these species, representing the genus *Paxillus*, differ (as do the Slimy Spike-caps overleaf) in several respects from other agarics. They are now placed alongside the boletes. The true agarics are discussed in vol. 2.

The Brown Roll-rim is very common and large numbers of these rust brown to olive brown fungi can be seen in mixed and deciduous woods. They are also frequently found in parks and gardens, on compost heaps and even on newly-laid lawns. This fungus often grows close to birch trees and probably forms mycorrhizal associations. The cap is first convex, with an in-rolled downy margin, then flat or depressed at the centre. It grows up to 12 cm broad. The dense, yellow-brown gills run downwards, and are forked close to the stem. As with boletus tubes, the gills of a Brown Roll-rim can be pulled away from the cap. The stem is tapering to start with, which is why the fungus resembles a wooden top. It can grow up to 6 cm tall. When bruised, as with *Leccinum aurantiacum*, it develops red-brown patches which soon darken. They are also thought to be closely related to *Gyrodon lividus*. Both the scent and taste are sourish. The Brown Roll-rim, which was thought to be edible, is in fact poisonous. Reports from central Europe indicate that raw or under-cooked specimens have caused serious cases of poisoning. Repeated consumption has even caused death. The growing season is from August to October.

Paxillus involutus has a yellow-brown cap and stem, but with Paxillus atrotomentosus *the stem is pale yellow and the cap almost white*

Paxillus atrotomentosus 🕸

Batsch ex Fr. (from Latin *atrotomentosus*: black-haired)

This Roll-rim is larger and more robust than the species described above. It grows on stumps or at the foot of living conifers, especially pines, and is easy to recognize by the short thick stem with a very dark brown velvety covering. Like the Brown Roll-rim it is not edible. Fairly common and widespread, growing from July to October.

Paxillus involutus

Paxillus atrotomentosus

Gomphidius glutinosus ⊖

Fr. (from Greek *gomphos*: plug, and Latin *glutinosus*: sticky)

Slimy Spike-cap

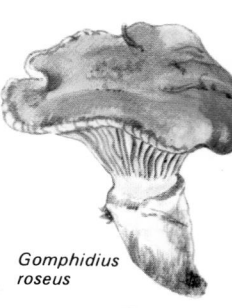

Gomphidius roseus

Easily recognized by the grey-brown slimy cap and brilliant yellow at the stem base. In Germany this fungus is sometimes called '*Kuhmaul*' ('cow snout'). The Spike-caps are thought to be close relatives of the genus *Suillus*, and similarly form mycorrhizal associations. The Slimy Spike-cap generally grows in fir woods and conifer plantations but not in native pine woods. The cap of the young fungus has an in-turned margin coated by a slimy veil which also covers the gills like a transparent membrane. In fully-grown specimens the cap is flat, and often develops dark patches which finally turn almost black. It grows up to 10 cm in width. The gills, widely spaced and running down to the stem, are first light grey, then dark with the black spores. The upper part of the stem is greyish white with a slimy ring often blackened by spores, a remnant of the veil. It grows up to 8 cm tall. The flesh is white but lemon yellow at the stem base. The scent and taste are both delicate. The Slimy Spike-cap is an edible mushroom, growing from July to November.

A closely related species is *Gomphidius roseus* ⊖ Fr. (from Latin *roseus*: rose-red). It is easy to recognize by the rosy colour of the cap, and by the fact that the cap does not develop black patches. The flesh is greyish white, slightly reddish at the base of the stem. It is thought to be specific to pine trees and is often associated with *Suillus bovinus* (see p. 114). The two species may be associated.

Chroogomphus rutilus ⊖

(Fr.) O. K. Miller (from Greek *chroma*: colour, and Latin *rutilus*: reddish)

Pine Spike-cap

Unlike the two Spike-caps described above, the Pine Spike-cap has no glutinous veil and is now therefore placed under another genus. This common fungus is confined to pine trees. The cap is pale to dark red-brown, sometimes wine-coloured. It is somewhat sticky, shiny in dry weather and up to 10 cm wide. The widely-spaced gills are the same colour as the cap. The stem is brownish yellow, tapers towards the base, and in young specimens has faint traces of a ring. It grows up to 10 cm tall. The flesh is reddish gold, but saffron at the base of the stem (see illustration on the right). The scent is slight and the flavour mild. The Pine Spike-cap is edible, and grows from July to November.

*Gomphidius
glutinosus*

Chroogomphus rutilus

Bibliography

General and reference works on fungi

Ainsworth, G. C., *Introduction to the History of Mycology*, Cambridge University Press, 1976.

Ainsworth, G. C. and Bisby, G. R., *A Dictionary of the Fungi*, Commonwealth Mycological Institute, 1971.

Ainsworth, G. C. and Sussman, A. S. (eds.), *The Fungi : An Advanced Treatise*, Vol. 1, *The Fungal Cell*, 1965, Vol. 2, *The Fungal Organism*, 1966, Vol. 3, *The Fungal Population Ecology*, 1968, New York and London, Academic Press.

Ainsworth, G. C., Sparrow, F. K. and Sussman, A. S. (eds.). *The Fungi: An Advanced Treatise*, Vol. 4A, *Ascomycetes and Fungi Imperfecti*, 1973, *Basidiomycetes and Lower Fungi*, 1974, New York and London, Academic Press.

Alexopoulos, C. J., *Introductory Mycology*, 2nd edn, New York, John Wiley & Sons, 1962.

Burnett, J. H., *Fundamentals of Mycology*, Edward Arnold, 1968.

Cartwright, K. St G. and Findlay, W. P. K., *Decay of Timber and Its Prevention*, 2nd edn, H.M.S.O., 1959.

Christiansen, C. M., *The Molds and Man. An Introduction to the Fungi*, 3rd edn, University of Minnesota, 1965.

Fries, E. M., *Systema Mycologicum*, Vols 1–2, Lundae, 1821–23, Vol. 3 and Index, Gryphiswaldae, 1829–32 (1821–32).

Haas, H., *The Young Specialist Looks at Fungi*, Burke Publishing Company, 1969.

Hawker, L. E., *Fungi. An Introduction*, 2nd edn, Hutchinson University Library, 1974.

Heim, R., *Champignons toxiques et hallucinogènes*, Paris, N. Boubée & Cie, 1963.

Ingold, C. T., *The Biology of Fungi*, 3rd edn, Hutchinson Educational, 1973.

Large, E. C., *The Advance of the Fungi*, Cape, 1940.

British fungus flora

Ainsworth, G. C. and Sampson, K., *The British Smut Fungi (Ustilaginales)*, Commonwealth Mycological Institute, 1950.

Barnett, H. L. and Hunter, B. B., *Illustrated Genera of Imperfect Fungi*, Minneapolis, Burgess Publishing Company, 1972.

Bourdot, H. and Galzin, A., *Hymenomycètes de France Hetérobasidiés – Homobasidiés gymnocarpes*, Bibliotheca Mycologica 23, Paris, J. Cramer, 1928.

Brightman, F. and Nicholson, B. E., *The Oxford Book of Flowerless Plants*, Oxford University Press, 1966.

Christiansen, M. P., *Danish Resupinate Fungi*, Part 1, *Ascomycetes and Heterobasidiomycetes*, Part 2, *Homobasidiomycetes*, Dansk Botanisk Arkiv 19, 1959, pp. 7–309.

Corner, E. J. H., *A Monograph of Clavaria and Allied Genera*, Oxford University Press, 1950; supplementary volume, *Beihefte zur Nova Hedwigia*, 33, 1970.

Dennis, R. W. G., British Ascomycetes, J. Cramer, 1968.

Dennis, R. W. G., Orton, P. D. and Hora, F. B., *New Check List of British Agarics and Boleti*, Supplement to *Transactions of the British Mycological Society* 43, 1960.

Domanski, S., *Fungi, Polyporaceae 1(resupinatae)*, *Mucronoporaceae 1 (resupinatae)*, translated from Polish; available from US Department of Commerce, Springfield, Virginia, 1972.

Domanski, S., Orlos, H. & Skirgiello, A., *Fungi, Polyporaceae II (pileatae)*, *Mucronoporaceae II (pileatae)*, *Ganodermataceae, Bondartzewiaceae, Boletopsidaceae, Fistulinaceae*, 1973; available as *Polyporaceae 1*.

Eckblad, F. E., *The Gasteromycetes of Norway*, Nytt magasin for botannikk, 1955, 19–86.

Ellis, E. A., *British Fungi*: Book 1, *Larger species*; Book 2, *Smaller Species*, Norwich, Jarrold Colour Publications, 1976.

Eriksson, J. and Ryvarden, L., *The Corticiaceae of North Europe*, Vol. 2 *Aleurodiscus to Confertobasidium*, Vol. 3, *Corticium to Hyphoderma*, 1975, Vol. 4, *Hyphodermella to Mycoacia*, 1976:

124

Vol. 1, with key to genera and glossary, will be published as a final volume: Norway, Fungi-flora.

Findlay, W. P. K., *Wayside and Woodland Fungi*, Frederick Warne & Co., 1967.

Grove, W. B., *British Stem- and Leaf-fungi* (*Coelomycetes*), Vol. 1, *Sphaeropsidales*, Vol. 2, *Sphaeropsidales and Melanconiales*, Cambridge University Press, 1935–37; reprint 1967, J. Cramer.

Henderson, D. M., Orton, P. D. and Watling, R., *British Fungus Flora. Introduction to Families and Genera*, Edinburgh, H.M.S.O., 1969.

Ing, B., *A Census Catalogue of British Myxomycetes*, The Foray Committee of the British Mycological Society, 1968.

Ingold, C. T., *An Illustrated Guide to Aquatic and Water-borne Hyphomycetes* (*Fungi Imperfecti*), Freshwater Biological Association, Scientific Publication No. 30, 1975.

Kuhner, R. and Romagnesi, H., *Flore analytique des champignons supérieurs*, Paris, Masson et Cie, 1953; reprint 1974.

Lange, M. and Hora, F. B., *Collins Guide to Mushrooms and Toadstools*, Collins, 1963.

Lister, A. and Lister G., *A Monograph of the Mycetozoa*, British Museum (Natural History), 1911.

Martin, G. W. and Alexopoulos, C. J., *The Myxomycetes*, University of Iowa Press, 1969.

Moser, M., *Kleine Kryptogomenflora*, (ed. H. Gams), Vol. 2a, *Ascomyceten*, Stuttgart, Gustav Fischer Verlag, 1963.

Moser, M., *Kleine Kryptogomenflora*, Vol. 2b, *Basidiomyceten II, Die Rohrlinge und Blatterpilze*, 3rd edn, Stuttgart, Gustav Fischer Verlag, 1967.

Pegler, D. N., *The Polypores, Bulletin of the British Mycological Society* 7: Supplement, 1973.

Ramsbottom, J., *A Handbook of the Larger British Fungi*, British Museum (Natural History), 1923; reprint 1965, J. Cramer.

Ramsbottom, J., *Mushrooms and Toadstools*, Collins, 1953.

Rayner, R. W., *Keys to the British Species of Russula, Bulletin of the British Mycological Society*, Vols 2–4, 1968–70.

Reid, D. A., *A Monograph of the British Dacrymycetales, Transactions of the British Mycological Society* 62, 1974, 433–94.

Richardson, M. J. and Watling, R., *Keys to Fungi on Dung, Bulletin of the British Mycological Society*, 2, 1968, 18–43; 2nd edn, 1975.

Rinaldi, A. and Tyndalo, V., *Mushrooms and Other Fungi – an Illustrated Guide*, Hamlyn, 1974.

Ryvarden, L., *The Polyporaceae of North Europe*, Vol. 1, *Albatrellus to Incrustoporia*, Norway, Fungiflora, 1975.

Singer, R., *Mushrooms and Truffles; Botany, Cultivation and Utilization*, Leonard Hill Books, 1961.

Singer, R., *The Agaricales in Modern Taxonomy*, 3rd edn, Vaduz, J. Cramer, 1975.

Wakefield, E. M. and Dennis, R. W. G., *Common British Fungi*, Gawthorn, 1950.

Watling, R., *British Fungus Flora 1. Boletaceae, Gomphidiaceae, Paxillaceae*, Edinburgh, H.M.S.O., 1970.

Watling, R., *Identification of the Larger Fungi*, Hulton Educational Publications, 1973.

Webster, J., *Introduction to Fungi*, Cambridge University Press, 1970.

Wilson, M. and Henderson, D. M., *British Rust Fungi*, Cambridge University Press, 1966.

For additional and more detailed references, the reader should consult the following list:

Holden, M., *Guide to the Literature for the Identification of British Fungi*, 3rd edn, *Bulletin of the British Mycological Society* 9, 1975, 67–106.

Index of common English names

Index of Latin names